Distributed Fiber Sensing and Dynamic
Rating of Power Cables

Distributed Fiber Sensing and Dynamic Rating of Power Cables

Sudhakar Cherukupalli and George J. Anders

IEEE
PRESS
SERIES
ON POWER
ENGINEERING

IEEE PRESS

WILEY

For general information on our other products and services or for technical support, please contact our Customer Care Department within the United States at (800) 762-2974, outside the United States at (317) 572-3993 or fax (317) 572-4002.

Wiley also publishes its books in a variety of electronic formats. Some content that appears in print may not be available in electronic formats. For more information about Wiley products, visit our web site at www.wiley.com.

Library of Congress Cataloging-in-Publication Data is available.

Hardback ISBN: 9781119487708

Cover design: Wiley
Cover images: (Top image) © Sam Robinson/Getty Images, (Center image) © asharkyu/Shutterstock

Set in 10/12pt Warnock by SPi Global, Pondicherry, India

V10014410_100819

This book would not have been possible without the love, sacrifice, and inspiration received throughout my life from my beloved mother Annapurna and father Gopalakrishna.

The second author would like to thank his wife Justyna for a continuous support and understanding and would like to dedicate this book to his grandchildren Sophia and Anthony Anders.

Contents

Preface

The subject of the book is the theory and practical applications of dynamic temperature sensing (DTS) in the context of high voltage (HV) power cables. The book is addressed to cable system design engineers, cable manufacturers, electric power system operators, engineering students, and scientists. This is the first book addressing a subject of the application of the DTS in HV power cables. DTS systems are used to obtain the temperature readings from the fiber optic sensors either built-in within the cable or placed on its surface or in close vicinity. Great majority of new HV cables are manufactured today with the fiber built-in. However, in order to take full advantage of this technology, the owner of the cable needs to familiarize himself/herself with the possibilities it offers. Hence, the book explains the physics of the DTS measurements and offers plenty of practical information about the costs, installation procedures, maintenance, and various applications – focusing on dynamic cable ratings.

The book is aimed as a primary source about the new area of temperature measurements for many different groups. The first group are cable manufacturers, who not only produce the cables with fiber optic links but also often offer the DTS and real time thermal rating (RTTR) systems themselves. The second group will be the DTS manufacturers. There are many companies around the world offering this technology. They usually understand the physics of the temperature sensing but have a limited knowledge of the utility practices, test and quality requirements, and possible rating applications.

The book is also addressed to the cable engineers. These will be utility personnel, contractors, and cable system designers. They usually understand the need for such systems, their output, and test requirements. However, they may lack the knowledge of the physics involved and the book will help them in understanding the opportunities and the limitations of the technology.

Another important group of readers will be comprised of the university students and their teachers. The book will help them appreciate the utility perspective in the application of the DTS technology. The book also has the classroom potential. It would particularly be useful for all the courses related to cable technology. This would include courses on cable construction, heat transfer phenomena in power cables, calculation of current ratings of electric power cables, and also design of ac and dc submarine cables that are being increasingly used to interconnect export from off-shore wind parks.

Considering the advantages that fiber optic sensors have to offer, it is not surprising their proliferation in a wide variety of modern industries. This book concentrates on the application of this technology in the electric power systems. It is organized as follows:

Chapter 1 contains introduction to fiber optic sensing with examples of several application fields not related to the power systems.

Chapter 2 provides a brief overview and talks about the distinction between single point and distributed measurements and discusses the advantages and disadvantages each of such systems offer.

Chapter 3 discusses the concept of distributed temperature sensing and describes what constitutes such a system as well as its architecture as used in the electrical power industry.

Chapter 4 introduces the reader to the optical fibers, connectors, optical cable construction, and provides some illustrations of these systems.

Chapter 5 provides examples of how optical fibers are incorporated into the land and submarine cables. It also discusses the advantages and disadvantages of various fiber locations inside and outside the power cable.

Chapter 6 discusses the DTS system requirements and reviews the standards for the electromagnetic compatibility. It presents the architecture of a DTS system and how it is integrated into a utility environment. Some of the challenges with this integration and data interpretation are also discussed.

Chapter 7 provides information of the importance of site testing of a DTS system and challenges that a utility may face with the site commissioning tests.

Chapter 8 discusses the relevance and importance of the DTS system calibrators. It also reviews the benefit of such systems and talks about the maintenance issues.

Chapter 9 illustrates how the temperature data may be utilized by the asset owner to optimize the use of their resources and discusses the dynamic circuit ratings of power cables with the data provided by a DTS system.

Chapter 10 Provides several examples of application of DTS systems in a utility environment. It includes a description of retrofitting of the fiber optic cables into existing cable circuits.

Chapter 11 provides some examples of potential future uses of the distributed fiber optical cable system for strain measurement and acoustic applications in power cables.

A large part of the material covered in this book was derived from various projects conducted by British Columbia Hydro (BCH) and from the work performed by various CIGRE and IEEE Working Groups of which the authors of this book were the members. The authors are indebted to Landry Molimbi and Masaharu Nakanishi, members of the CIGRE WG B1.45, who provided background material used in Chapters 4, 7, and 8 of this book. Frequent discussions with DTS vendors contributed greatly to the development of many procedures described herein. In addition, we could have not written this book without an involvement and close association with several individuals who contributed their ideas and took the time to read the manuscript. We are particularly indebted to Dr. M. Ramamoorty and Chris Grodzinski, who have reviewed the entire manuscript and provided several helpful comments.

We would also like to acknowledge the support the authors have received from the Standards Council of Canada (SCC) and several Senior Managers from BCH, as well as the financial assistance of SCC and BCH in supporting our participation in the activities of WG 10 of the IEC and WG B1.45 of CIGRE over many years.

Finally, but by no means last, we would like to thank our families who supported wholeheartedly this endeavor.

Vancouver and Toronto   *Sudhakar Cherukupalli*
                                                     George J. Anders

Acknowledgments

Dr. Sudhakar Cherukupalli thanks Management at BC Hydro for allowing the generous use of photos used in many chapters that were taken during several projects executed during the author's association with these projects. Material for this book has been collected through discussions, correspondence, and support from several DTS vendors who were willing to share their current state of knowledge and development in their respective organizations. The challenges posed by the author to many of the DTS vendors helped them to improve and refine their products which have led to great interest in these distributed sensing technologies and apply them in the power industry today.

Dr. Sudhakar Cherukupalli would like to acknowledge retired colleagues from BC Hydro and in particular Joseph Jue, Allen MacPhail, and Takashi Kojima for their support during the world's first installation to harness DTS technology on a 525 kV submarine cable system.

1

Application of Fiber Optic Sensing

Today, fiber optic (FO) sensors are used to monitor large composite aircraft structures, concrete constructions, and to measure currents in high-voltage equipment. They are also applied in electrical power industry to measure electric fields as an alternate to current and voltage transformers. They are also finding many applications in the field of medicine, chemical sensing, as well as to monitor temperatures around large vessels in the oil and gas processing industries. There have been recent attempts in Japan to monitor the wings of a fighter aircraft to monitor dynamic strain and temperatures when the aircraft is taking off and landing to better understand the load and temperature-induced stresses and how these affect fatigue performance. Considerable work is underway to map strain in large composite and concrete structures.

Research on the application of FO sensors (FOSs) has been conducted over many years. They were first demonstrated in the early 1970s (Culshaw and Dakin 1996; Grattan and Meggitt 1998) and are the subject of considerable research since 1980s. Early applications were focused on military and aerospace uses. FO gyroscopes and acoustic sensors are examples, and they are widely used today. With the increase in the popularity of FOSs in the 1980s, a great deal of effort was made toward their commercialization with an emphasis on the intensity-based sensors. In the 1990s, new technologies emerged, such as in-fiber Bragg grating (FBG) sensors (Morey et al. 1989; Rao 1997), low-coherence interferometric devices (Grattan and Meggitt 1995; Rao and Jackson 1996), and Brillouin scattering distributed sensors (Bao et al. 1995). Dramatic advances in the field of FO sensors have been made as a result of the emergence of these new technologies leading to a significant proliferation of their use.

FO sensors offer significant advantages over conventional measuring devices, most important of which are: electromagnetic immunity (EMI), small size, good corrosion resistance, and ultimate long-term reliability. For example, FBG sensors offer a number of distinguishing advantages including direct absolute measurement, low cost, and unique wavelength-multiplexing capability.

Distributed Fiber Sensing and Dynamic Rating of Power Cables, First Edition.
Sudhakar Cherukupalli and George J. Anders.
© 2020 by The Institute of Electrical and Electronics Engineers, Inc.
Published 2020 by John Wiley & Sons, Inc.

These new measuring technologies have formed an entirely new generation of sensors offering many important measurement opportunities and great potential for diverse applications.

This book is devoted to the application of the FO technology in electric power cables. However, before we tackle this subject, we would like to offer a brief review of the application of this technology in other fields. In the following sections, an attempt is made to provide an overview of the types of FO sensors and we will list some of the industries where FOSs have been applied. These include:

- Oil and gas industry
- Fire detection and integrating them with the firefighting equipment
- Large composite structure such as bridges and dams
- Mines
- Aircraft industry
- Medicine
- Power industry

1.1 Types of Available FO Sensors

One of the first applications of FOSs involved the, so-called, limit sensors with an ability to detect motion beyond a certain limit and initiate action once this limit is exceeded. These types of sensors can be used for monitoring linear or rotational motion. Another type of an early application is a level sensor when a solid or a liquid rises or falls beyond a set point. Proximity sensors use infrared emission, reflection, or pressure change to perform such detection without the need for any physical contact. Another example involves a beam of light crossing a doorway. Beam interruption can be detected by a photo sensor and trigger an alarm. This application is typically used in the process industry for counting or having access control.

Another class of sensors uses FOs for linear or angular position control. One may have an array of optical fibers that are placed in parallel. The object to be detected when passing this array of sensors may alter the transmission or reflection of light. The sensor processing electronics can then infer the proper position of each object within the sensing region. In this case, the resolution of the detected positions will depend on the spacing between the sensing points. This idea has been extended by placing fibers in an angular fashion and has been used to detect or establish the angular position of a gear on drive systems.

FO sensors have also been used to measure linear as well angular speed or velocity of shafts (tachometers). Some of them use Doppler phase shift methods for such measurements.

A FO gyroscope is another type of sensor. It consists of a coil, either polarization preserving or not, of optical fibers in which the light is simultaneously propagating in the clockwise and counterclockwise directions. The SAGNAC effect induces a differential phase shift between the clockwise and counterclockwise guided waves in the rotating media. The phase difference in the detected signal is converted into a rate or angle of rotation. Examples of this application are the Brillouin or resonant FO gyroscopes.

Optical fibers have also been used for temperature sensing. They may be broadly classified as FBG devices. Phosphor coated, Fabry–Perot cavity terminated, or thermo-chromatic terminated optical fibers are some examples of such point sensors. In the electric power industry, these devices have been used to measure transformer winding temperatures.

Another class of temperature sensors are the distributed FO temperature devices. These are broadly classified as based either on the Raman scattering or Brillouin principles. Laser light injected into a fiber is continuously scattered. This backscattered light is used for calculating temperature profiles along the fiber. Brillouin scattering exhibits sensitivity to temperature as well as strain, so care has to be exercised when interpreting the temperature data. Devices based on Raman scattering fall into two principal categories: those that rely on the optical time domain reflectometry (OTDR) while the others use the optical frequency domain reflectometry (OFDR). It is the latter that provides the highest spatial resolution along a fiber.

While discrete monitoring based on FBG FO system has been and is continued to be used, it is expected that the distributed FOSs will play a more important role in health monitoring of large structures, such as dams, because the use of a single fiber makes installation easy with the ability to measure over long distances. While considerable advances have been made in measuring strain, the effects of temperature-induced strain appear elusive especially when one needs to improve spatial resolution and accuracy. Because of its complexity, simultaneous measurement of multi-axis strain and temperature in composites remains a major challenge for the optical fiber sensors community.

Because of a spate of explosive in-service failures of the 525 and 800 kV oil-filled current transformers in the late 1990s, the industry was looking for alternate devices that could provide accurate current measurements with a dynamic rating ranging from several 1000s to several 100 000s amperes. This led to the development of FO electric current sensors based on the Faraday effect. After more than two decades of development, they have successfully entered the market. At the University of British Columbia in Canada, Dr. Jaeger and his research team developed an electric field or voltage sensors and had conducted several field tests to prove the concept and develop a practical system. These systems were placed in the local utility's 500 kV substation on a trial basis (Jaeger et al. 1998). Further collaborative efforts led to the development of integrated electro-optic voltage and current sensors. Initially, the challenge with these systems appeared to be their inability to match the needs of the conventional protection equipment that required 120 V and 5 A input signals. In parallel, the modernization of the electric power grid and changes that occurred with the development for solid-state relays, which now require lower voltage (~5 V) and current signals (100 mA), the devices developed at the university became more appealing to the industry. Moreover, these electro-optical devices provided higher sensitivity, improved frequency bandwidth, with significantly improved EMI. This led to general acceptance of electro-optical devices for current and voltage measurements in the power industry.

For medical applications, extrinsic FOSs based on multimode fiber transmission have matured. A number of FBG temperature and ultrasound sensor systems have been developed. With further engineering, it is anticipated that these systems could be used for in-vivo measurement of temperature or/and ultrasound as well as blood pressure in the heart and other organs. In chemical sensing, FOSs based on evanescent wave coupling is still under development.

The following sections describe in more detail the application of the FO sensors in various industries.

1.2 Fiber Optic Applications for Monitoring of Concrete Structures

When compared with traditional electrical gauges used for strain monitoring of large composite or concrete structures, FOSs have several distinguishing advantages, including:

- A longer lifetime, which could probably be used throughout the working lifetime of the structure (e.g. >25 years) as optical fibers are reliable for long-term operation over periods greater than 25 years without degradation in performance.
- A greater capacity of multiplexing a large number of sensors for strain mapping along a single fiber link, unlike strain gauges, which need a huge amount of wiring.
- Greater resistance to corrosion when used in open structures, such as bridges and dams.
- A much less intrusive size (typically 125 mm in diameter – the ideal size for embedding into composites without introducing any significant perturbation to the characteristics of the structure).
- A much better invulnerability to electromagnetic interference, including storms, and the potential capability of surviving in harsh environments, such as that encountered in the nuclear power plants (Townsend and Taylor 1996).

These features have made FOSs very attractive for quality control during construction, health monitoring after building, and impact monitoring of large composite or concrete structures (Udd 1995). Since the use of FOSs in concrete was first suggested in 1989 (Mendez et al. 1989) and the demonstration of embedding a FO strain sensor in an epoxy–fiber composite material was reported in the same year (Measures 1989), a number of applications of FOSs in bridges, dams, mines, marine vehicles, and aircraft have been demonstrated.

One of the first monitoring demonstrations for large structures using FOSs was a highway bridge using carbon fiber-based composite prestressing tendons for replacement of steel-based tendons to solve a serious corrosion problem (Measures et al. 1994). Because composite materials are not well proven in their substitution for steel in concrete structures, there is considerable interest in monitoring the strain and deformation or deflection, temperature, or environmental degradation within these materials using an integrated FO sensing system. FBG sensors could be suitable for achieving such a goal. An array of FBGs has been attached to the surface of a composite tendon and the specially protected lead-in/out optical fibers egress through recessed ports in the side of the concrete girders, as shown in Figure 1.1. However, if the FBG sensors could be embedded into the composite tendons during their manufacture, excellent protection for the sensors and their leads would be provided and done by (Measures et al. 1997).

Dams are probably the biggest structures in civil engineering; hence, it is vital to monitor their mechanical properties during or/and after construction in order to ensure the construction quality, longevity, and safety. FOSs are ideal for health monitoring applications of dams because of their excellent ability to be used in long-range measurements. Truly distributed FOSs are particularly attractive as they normally have tens of kilometers measurement range with a meter spatial resolution. A distributed temperature sensor has been demonstrated for monitoring concrete setting temperatures of a large dam in Switzerland (Thevenaz et al. 1999). This monitoring is of prime importance as the density and microcracks are directly related to the maximum temperature the concrete experiences during the setting chemical process.

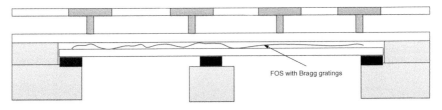

FOS with Bragg gratings

Figure 1.1 Schematic diagram of a fiber Bragg grating sensor locations for strain monitoring on a bridge.

This system has been used for concrete setting temperature distribution in a slab with dimensions of $15\,\text{m}\ (L) \times 10\,\text{m}\ (W) \times 3\,\text{m}\ (H)$. These slabs are used for estimating the height of the dam in order to increase the power capability of the associated hydroelectric plant. The layout of an optical communication cable inside the slab is shown in Figure 1.2, which gives a two-dimensional temperature distribution of the whole area. The fiber cable is installed during the concrete pouring. Figure 1.3 shows the temperature distribution over the slab at different times after concreting. It reveals that the temperature at the central area of the slab can be as high as $50\,°\text{C}$ and it takes many weeks for this region to cool down.

Many research groups have demonstrated the "simple" operation of FBGs embedded in large composite or concrete structures for strain measurement. However, further work may be needed in order to produce a cost-effective, multifunctional FBG multiplexing system that is able to measure static strain, temperature, and dynamic strain simultaneously with adequate resolution and accuracy. With such system, the FBG sensor would be able to realize its full potential and perhaps dominate the market for health monitoring of large composite and concrete structures in the future.

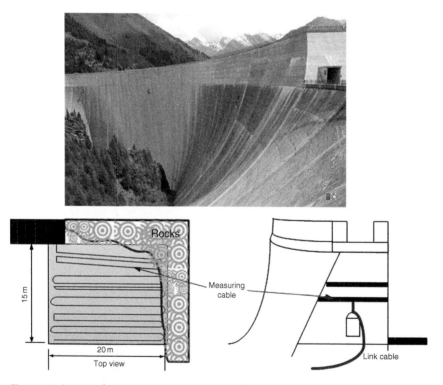

Figure 1.2 Layout of an optical communication cable inside a concrete slab.

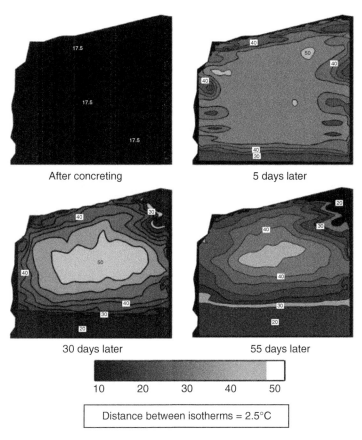

After concreting   5 days later

30 days later   55 days later

10   20   30   40   50

Distance between isotherms = 2.5°C

Figure 1.3 Two-dimensional temperature distribution in Luzzone dam during the concrete curing obtained with embedded fiber using Brillouin sensor.

1.3 Application of FO Sensing Systems in Mines

Measurement of load and displacement changes in underground excavations of mines and tunnels is vital for safety monitoring. Multiplexed FBG sensor systems could replace the traditional electrical devices, such as strain gauges and load cells, which cannot be operated in a simple multiplexed fashion and in a very hazardous environment with strong electromagnetic interference generated by excavating machinery. An FBG sensor system based on a broadband Erbium-doped fiber source and a tunable Fabry–Perot filter has been designed for long-term static displacement measurement in the ultimate roof of the mining excavations and in the hanging wall of the ore body's mineshaft (Ferdinand et al. 1995). A specially designed extensometer with a mechanical-level mechanism can cope with the large displacements of up to a few centimeters by controlling the overall strain change of the FBG to be less than 1%. This system was also field tested by the authors.

1.4 Composite Aircraft Wing Monitoring

Advanced composite materials are now routinely used for manufacturing aerospace structures (e.g. parts of airplane wings). Compared with metallic materials, advanced composite structures can have higher fatigue resistance, lighter weight, higher strength-to-weight ratio, the capability of obtaining complex shapes, and no corrosion. Hence, the use of composite materials with embedded FBG systems can lead to a reduction in weight, inspection intervals, and maintenance cost of an aircraft – and, consequently, to an improvement in performance (Foote 1994). However, there is a major challenge in analyzing the real-time health and usage monitoring with an onboard sensor system. A distributed FBG sensor system could be ideally suitable for such an application. Because FBG sensors are sensitive to both strain and temperature, it is essential to measure strain and temperature simultaneously in order to correct the thermally induced strain for static strain measurement. A simple and effective method often used is to employ an unstrained temperature reference FBG, but this approach is not suitable for FBG sensors embedded in composites. A number of approaches have been proposed for the simultaneous measurement of strain and temperature (Jones 1997; Liu et al. 1997; Rao 1997), but there are still some issues that have to be resolved. These include:

- How can one integrate both strain and temperature sensors into the same location within the composite to obtain high spatial resolution?
- How to achieve adequate strain–temperature discrimination accuracy with the dual-parameter method and especially if multiplexing is also needed?
- How to process the distorted FBG signal if the FBG experiences a nonuniform strain along its length – this can occur when the applied strain is large?

Optical fiber sensors offer a number of advantages for spacecraft applications. A principal application is strain sensing for structural health monitoring, shape determination, and spacecraft qualification testing. The Naval Research Laboratory has examined how optical fiber strain sensors have been used on spacecraft structures and ground test hardware. The sensors have been both surface mounted to the structure and embedded in the fiber-reinforced polymer composites. The issue of potential strength reduction of high-performance composites due to embedded optical fiber sensors and leads has been studied, low-cost fabrication of tubular struts with embedded sensors has been demonstrated, and a novel technique for fiber ingress–egress from composite parts has been developed. Friebele et al. (1994) also discuss the applications of fiber sensors, distributed dynamic strain monitoring of a honeycomb composite plate, and a lightweight reflector system during acoustic qualification tests, ultrahigh-sensitivity static strain and temperature measurements for precision structures, and online identification of a lightweight laboratory truss.

1.5 Application in the Field of Medicine

The majority of commercial sensors widely used in medicine are electrically active and hence not appropriate for use in a number of medical applications, in particular, in high microwave, radiofrequency, or ultrasound fields. Another example is laser radiation associated with hyperthermia treatment. Because of local heating of the sensor head and the surrounding tumor due to the presence of metallic conductors and electromagnetic interference of currents and voltages in the metallic conductors, the readings are often burdened with a significant error. FO sensors can overcome these problems as they are virtually dielectric. A range of miniature FO sensors based on intensity modulation has been successfully commercialized (Trimble 1993). Generally speaking, these are all point sensors that can only provide readings over a small volume in the human body. Although their passive multiplexing is possible, it is difficult to achieve in practice because of the limitations on the probe size. By using the unique multiplexing property of the FBG sensor, it is possible to make quasi-distributed sensor systems with a single fiber link. A number of temperature and ultrasound sensing systems have been demonstrated to date.

Similar to temperature monitoring for the assessment of the operating safety of high-RF or microwave fields mentioned above, an ultrasound sensor is required to monitor the output power from the diagnostic ultrasound equipment used for a range of medical applications (including ultrasound surgery, hyperthermia, and lithotripsy). Piezoelectric (PZ) devices are the most common sensors but suffer from susceptibility to electromagnetic interference and signal distortion and from the difficulty of direct determination of ultrasound fields in vivo because of the limitation of the probe size. The FBG sensor can overcome these problems and is able to measure the ultrasound field at several points simultaneously thanks to its unique multiplexing capability.

1.6 Application in the Power Industry

Electric current measurement using FOSs in 1977 is probably the earliest application of this technology in the electric power industry (Rogers 1988). After more than two decades of development, fiber-optic current sensors have entered the market (Bosselmann 1997). These current sensors based on the Faraday effect have found important applications in fault detection and metering, and there is already a great deal of literature dealing with this subject. In this chapter, we focus on some recent progress in applications of FBG sensors in the electrical power industry with details concerning electric power cables discussed in later chapters of this book.

Like other implementations of FO sensors, the FBG is ideal for use in the electrical power industry because of its immunity to electromagnetic interference. In addition, the FBG can be written onto standard 1550-nm-wavelength telecommunication fiber; hence, long-distance remote operation is feasible because of low transmission loss. Loading of power transmission lines, winding temperature of electrical power transformers, and large electrical currents have been measured with the FBG sensor.

We will start by providing a brief literature review on the applications of the FO cables in the electric power industry.

1.6.1   Brief Literature Review

There is a considerable amount of published literature about distributed temperature sensing (DTS) technology in the power industry. This section looks at important publications concerning power system applications with the emphasis on the real-time conductor temperature and cable ampacity calculations. The following material is based on the work of CIGRE WG B1.45 of which the first author of this book was a member.

A 1987 paper presented by an engineer from EDF (Duchateauf 1987) explored the usage of transferring data using FO cables along a 20 kV cable circuit. The utility studied how an optical fiber could be placed alongside a power cable and with what configurations, how joints between sections can be made, and what was the impact on the cable installation costs. They determined that installation of optical sensor alongside power cable is possible, irrespective of the laying method. The author concluded that the additional cost for laying of the optical fiber was negligible if done when the main power cable was being installed.

Marsh et al. (1987) discussed the opportunities provided by the boom in the offshore oil and gas industry and the benefits achieved leading to the development of composite submarine cables incorporating power, data control, and telecommunications cores. The authors discuss how these cables had been designed to meet the specific requirements of a platform to platform or mainland to platform links. They also discussed how advancements in the optic fiber technology was able to satisfy the requirements for high-capacity data transmission links, which, in turn, led to developments of composite power and optic fiber cables.

Cable manufacturers (Cantergrid et al. 1987) also discussed the challenges of composite submarine cables (power and optical fibers). Manufacturers developed test programs to be conducted in their laboratories with a goal of ensuring good on-site performance.

As early as in 1987, Pirelli engineers recognized that optical sensors embedded in power cables are invaluable for communication purposes (Giussani et al. 1987). They also pointed out to several design challenges when incorporating such sensors into the submarine power cables. In a submarine power cable with integrated optical fibers, the optical sensor will be exposed to high hydrogen content, as this gas can be generated by very active electrolytic cells like the Zn—Fe in the zinc-coated cable armor, or by the cathodic protection of the metallic structures immersed in water as in the case of off-the shore platforms. This can cause serious attenuation of the measured signals (hydrogen peak). They indicated that contact between H_2 and fibers must be avoided. This condition implied that the optical fiber must be protected by a continuous metal sheath. The extruded lead alloys and longitudinal welded copper sheaths have been used with success.

In 1998, Donazzi and Gaspari (1998) reported the development of the real time thermal rating (RTTR) software used on a field trial on a multi-zone installation in Italy allowing continuous monitoring of a cable circuit with prediction features. Upon successful conclusion of this field trial, first application on a utility's HV cable system was implemented in 1998 in the UK (Pragnell et al. 1999), which ran for a one-year period. The installation involved a 132 kV buried cable system that was only 500 m long but incorporated 18 thermal zones within the RTTR program. The circuit was installed in 1993 and had optical fiber sensors attached to the cable surface. Each thermal zone was considered separately within the model. The temperatures measured using the DTS were compared with those calculated using assumed thermal resistances for the environment. The values were then adjusted such that the model temperatures matched the actual DTS measurements.

Balog et al. (1999) describe the performance of DTS and RTTR systems used to monitor the ratings along a submarine cable in the shore section with the optical fiber attached to the submarine cable surface. They discussed how the economics for deployment of such a system will depend upon the value of additional energy that may be transmitted by the power cable. The trade-offs will be between the avoided penalty for nondelivery of power and the value of the delayed investment, versus the investment and operating cost of the DTS/RTTR systems.

Another DTS and RTTR system was implemented under a test program on a 400 kV installation in France that included various methods of cable installation as well as different types of temperature sensors (Luton et al. 2003). The authors reported different operating scenarios (steady-state and transient overloading) and the resulting ampacities were evaluated. They also discussed the outlook for the application of such monitoring technologies on cable systems.

Su et al. (2003) discuss how DTS measurements in conjunction with the finite element analysis can be used to determine thermal resistivity and ground temperature at individual hot spots along a cable corridor.

Engineers at Hydro Quebec (Leduc et al. 2003) implemented a DTS system to monitor several 120 kV cable circuits and performed rating calculations with the finite element analysis. Working with three parameters: time, ampacity and temperature, the software computed the value of one of them with given values of the two other variables.

A retrofit installation of an optical fiber in the hollow conductor of an operating 525 kV Ac submarine cable system in Canada to allow DTS measurements is described by Cherukupalli (2006) and Cherukupalli et al. (2018). The system is now used for managing and optimizing the power transfer over the cable circuit.

Popovac and Damlianovic (2006) describe how thermal monitoring along a 110 kV cable in Belgrade has made it possible to identify new hot spots along the route and for the dispatcher having information at hand for allowing short-term overloads.

Smit and de Wild (2006) present a dynamic rating system for an existing 150 kV transmission circuit in the Netherlands, consisting of both cables and overhead (OH) lines. Based on the present conditions (as known from load history, ambient conditions, and thermal models), the system calculates the possible overload capacity of the line. The system, which is based on thermal models rather than on measurements, has operated satisfactorily since the spring of 2005.

After a description of the encountered experiences, Avila and Vogelsang (2010) focus on the added value generated by the advanced diagnostic monitoring systems (DTS and partial discharge monitoring) of the cable link and the operation of the cable system. Their paper describes the design, production, testing, and installation of a 500 kV cable system including the monitoring equipment at the Sidi Krir Power Station (750 MW) near Alexandria in Egypt and addresses the following aspects: design, manufacturing, routine testing of the cables as well as the installation of the cable system on site.

Installation of a DTS and a real-time rating system to monitor a 345 kV SCFF cable system in a tunnel is reported in the work of Lee et al. (2003). An unusual issue discussed in this paper relates to the observation that over the period of a day the "hot spot" moved over a distance of 70 m requiring careful analyses and prediction of the ratings.

The possibilities of an online dynamic rating system was demonstrated on a pilot project at the NUON utility (de Wild et al. 2007) in which a 150 kV connection consisting of an oil-filled power cable and an OH line in series was set up. The goal was to find thermal bottlenecks in the cable circuits paying attention to the relation between thermal models and measurements.

Most long high-voltage subsea cable installations are challenging and as a result there are risks that have to be mitigated. Some, like scour and sediment, can be unpredictable with sometimes a few years of no change followed by a changing pattern of the seabed movement. These issues can lead to premature cable failures. Kluth et al. (2007) introduce the concept of dynamic temperature and strain sensing and describe how this can be related to cable bending and stretching. They further describe how this can be effectively used to interpret stresses and strains experienced by a submarine cable. Because of this ability to measure both distributed temperature and strain sensing (DTSS) with a standard embedded fiber, the authors suggest it can be used to monitor thermal and mechanical problems before any premature cable failure can occur.

Accurate computation of the real-time thermal rating of cables has the fundamental problem that not all parameters, such as the material properties and geometry, are well known along the cable route. They may change along the route with time. Brakelmann et al. (2007) developed a RTTR system which was able to predict heating trends and/or load reserve margins using measured temperatures from FO monitoring systems. They also showed how critical system conditions can be identified at an early stage and alarm messages can be sent to the operator. Some attention was devoted to parameter adaptation by introducing a unique algorithm, which proved to be accurate and extremely robust. They subsequently performed tests for extreme initial situations, such as cable in close proximity to a neighborhood district heating pipes and other cables, which were not initialized in the thermal model, and soil dry-out condition. It was noted that such scenarios lead to extremely bad initial values which had to be adapted in the model and analyzed. They demonstrated that even under such poor (extreme) initial conditions, the forecast errors normally remain below a limit of 1 K.

In early 2000s, a cable monitoring system was installed on a submarine HVDC link between Sweden and Germany with the goal of increasing ampacity of this 220 kV circuit. The utility wanted to have a full utilization of the transmission capacity of the Baltic Cable. Compared to the initial ampacity assessed with conservative assumptions, the current rating of the cable was optimized by measuring and evaluating the conductor temperature resulting in a 12% increase in cable ampacity compared with the design value (Schmale and Drager 2010). Calculation of a dynamic current rating directly in the control center was considered as a possible next step.

Paper by Domingues (2010) gives an insight into a dynamic current capacity assessment system, employing distributed thermal measurements by integrated optical fiber. This system was implemented in a 1.1 km 138 kV cable circuit in Brazil. The paper addresses the thermal modeling as well as the compilation and the analysis of the measurement data.

Prompted by the advancements being made in the "intelligent" power system equipment, Kim et al. (2011) describe how they used optical fibers by embedding them in the distribution class cables to provide partial discharge (PD) data from cable splices as well cable temperatures along the circuit corridor.

Liu and Wang (2011) discuss the importance of fiber splicing on the accuracy of the DTS measurement.

Jacobsen et al. (2011) describe an application of the dynamic rating systems on a 145 kV cable circuit that transfers power from the Rødstand offshore windfarm in Denmark. The cables had both submarine and land sections. An early analysis of the recorded temperatures has shown poor soil conditions due to ancient organic material and this happened despite remedial steps taken during the installation. In contrast, the land section was not the thermally limiting portion of the route even though the analysis performed during the design stage would suggest so.

A utility in China (Yan et al. 2011) reported an unusual event and described how an OFDR-based DTS system helped them to identify and resolve a problem on their 220 kV cable network. The authors found water had leaked into a cross-bonding box in a manhole. This led to a short circuit inside the box resulting in water temperature reaching 84 °C triggering an alarm and transmittal of the SMS text messages to the utility engineers about the situation. The DTS system measuring the temperatures in the manhole showed a dramatic increase which helped them locate, identify, and resolve the problem.

Engineers at Tennet (Schmale et al. 2011) installed, on an exploratory basis, an integrated DTS and RTTR system to dynamically rate their three-phase 220 kV XLPE cable placed in a steel pipe. They found by using the DTS system potential static rating increase in the order of 12%. Early results with the RTTR systems showed that there was considerable more potential for an increase in the power flow. Their study was limited to the winter season and reported that further work had to be done to study, analyze, and then finalize the ratings following another season of monitoring.

A utility in the United States was replacing their high pressure fluid-filled cables with three-phase XLPE designs (Igi et al. 2011). They placed a FO cable in the interstitial space and in a communication duct located on top of the pipes. The authors describe how they monitored the fiber temperature in the interstitial space as well as in the communication duct using two DTS systems. To confirm the accuracy of their thermal model and the equivalent thermal resistance between the communication duct and the power cables, it was necessary to compare both temperatures for an extended period of time so that the temperature variation in the transient condition or daily load cycle could be evaluated. The ultimate goals were to relate the interstitial temperature with the temperature of the fiber outside the steel pipe by altering/adjusting the thermal resistivities. Once this correlation match was obtained, it was not difficult to relate the interstitial and the cable conductor temperatures and thereby provide accurate steady-state and transient ratings for this corridor. In this particular instance, the DTS system was placed in a manhole and powered by batteries for a short period of time.

The experience from Belgium on PD and temperature monitoring techniques aimed at establishing the condition of the cable is described in the work of Hennuy et al. (2014). The future introduction of DTS monitoring systems for real-time follow-up and the use of optical fiber in directional drillings are also mentioned in order to get the maximum possible loading of the cable system.

A pilot project for determining the optimal current-carrying capacity during operation and to find thermal bottlenecks in time is presented in the work of Chimi et al. (2014). The measured results were compared with the values computed with a FEM-based software module. These results show good correlation and the conclusion is that the cable temperature monitoring can be a valuable tool for future grid planning.

Experience gained using a dynamic rating system is reported in the work of Jacobsen and Nielsen (2012). The authors describe a load management system installed on four 145 kV lines including submarine and land portions. The description includes the system setup and its integration into the supervisory control and data acquisition (SCADA) system. One of the conclusions reached by the authors is that, during the first years of operation, cables had a slightly higher current rating than the design parameters would have suggested. In addition, a hot spot was detected, which was subsequently investigated.

Calculations of transmission capacity at the design stage are based on several assumptions. Temperature measurement and calculation of actual transmission capacity are discussed in the work of Wald (2012), where the principles of measurements with DTS are described and illustrated by a practical example. The case study concludes that calculations performed according to the IEC standard with an assumption of nominal environmental parameters may be conservative. Use of temperature measurements, collection of more precise site data, and recalculation with a finite element method allowed a better understanding of the behavior of the cable link.

DTS is increasingly being used for hot spot detection. Coupling the DTS measurements with some advanced mathematical models to estimate the conductor temperature enables dynamic rating of the cable system. This gives many benefits for the system operator but requires knowledge of the dynamic thermal behavior of the surroundings of the cable. Two such dynamic systems have been installed in the SEAS-NVE's grid and described in the work of Jacobsen et al. (2012). Other systems are planned.

1.6.2 Monitoring of Strain in the Overhead Conductor of Transmission Lines

An excessive mechanical load on electrical power transmission lines, which may be caused, for example, by heavy snow, may lead to a serious accident. In particular, for those lines located in mountainous areas, there is no easy access for inspection. Therefore, an online measurement system is needed to monitor the changing load on the power line. A multiplexed FBG system with more than 10 sensors distributed over a distance of 3 km has been demonstrated, as

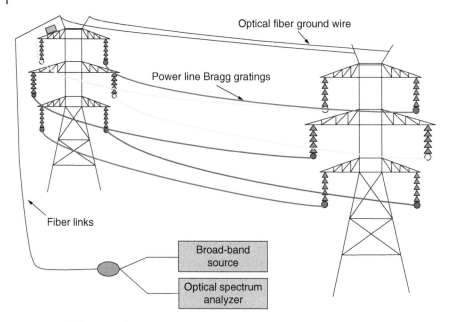

Figure 1.4 Schematic of FBG load monitoring system for power transmission lines (Ogawa et al. 1997).

shown in Figure 1.4 (Ogawa et al. 1997). The load change is simply converted into strain via a metal plate attached to the line and onto which the FBG is bonded. Obviously, many more sensors are required for such application. Wave Division Multiplexing may no longer be able to cope with the significant increase in the number of sensors because of the limited bandwidth of the light source. However, time division multiplexing could be used to considerably improve the multiplexing capacity. As the distance between adjacent FBGs is large, high-speed modulation and demodulation would not be required. Overall, this is an excellent example of applying FBG sensors for long-distance remote monitoring in harsh environments.

1.6.3 Temperature Monitoring of Transformers

Knowledge of the local temperature distribution present in high-voltage, high-power equipment, such as generators and transformers, is essential in understanding their operation and in verifying new or modified products. Defective or degraded equipment can be detected by continuously monitoring the variations in the winding (the "hot spot") temperature, which reflects the performance of the cooling system. The FBG sensor has been demonstrated for such an application where the winding temperature of a high-voltage transformer is measured with two 1550 nm FBGs with sensing and reference elements interrogated via a standard optical spectrum analyzer (Hammon and Stokes 1996). A measurement accuracy of ±3 °C had been achieved for long-term monitoring.

1.6.4 Optical Current Measurements

Optical fiber sensors exploiting the Faraday effect have been intensively researched and developed for measurement of large currents at high voltages in the power distribution industry for more than a decade (Day et al. 1992). However, problems associated with induced linear birefringence,[1] temperature, and vibration have limited the application of this technique. An alternative method is to measure the large current indirectly by using a hybrid system consisting of a conventional current transformer (CT) and a PZ element. The CT converts the current change into a voltage variation, and then this voltage change is detected by measuring the deformation of the PZ using an FBG sensor (Henderson et al. 1997). An interferometric wavelength-shift detection method has been exploited for the detection of wavelength shift induced by the current and a current resolution of 0.7 A/pHz over a range of up to 700 A has been obtained with good linearity. More recently, the resolution has been further improved by replacement of the FBG with an FBG-based Fabry–Perot interferometer (FFPI) formed by arranging two FBGs with the same central wavelengths along a length of the fiber. Figure 1.5 (Rao et al. 1998) shows the experimental results for measurement of a 50 Hz, 12 A current and the performance of the sensor, respectively. From Figure 1.5b, one can see that good linearity has been obtained. Compared with expensive CTs used by the industry, these approaches offer a much lower cost alternative as the sophisticated electrical insulation is no longer required.

1.7 Application for Oil, Gas, and Transportation Sectors

In 1997, LIOS Technology GmbH introduced "fiber optical linear heat detection" for efficient and pinpoint fire detection in road and rail tunnels. With more than 3500 installations worldwide, this has revolutionized the safety standard in underground transport facilities as, for the first time, it enabled the complete monitoring of the entire tunnel length. The line sensor is based on optical waveguides, which are immune to electromagnetic disturbances. Based on years of experience within the field of fiber-optic measuring techniques as well as the input of research partners and customers, multiple characteristics

---

1 "Birefringence" is the optical property of a material having a refractive index that depends on the polarization and propagation direction of light. Some optically anisotropic materials are said to be "birefringent."

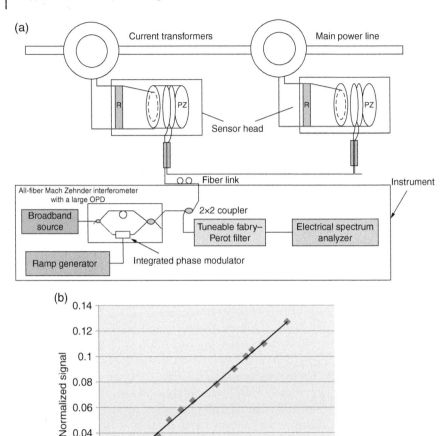

Figure 1.5 Results of current measurement using FBG-based Fabry–Perot interferometer. (a) Schematic diagram of the multiplexed FBG-based Fabry–Perot current sensor (Rao et al. 1998). (b) Normalized sideband signal amplitude as a function of effective current at 50 Hz (Rao et al. 1998).

were added to offer reliable and adaptable measurement features for locally linear heat detection profiles. Since this is a purely passive sensor, unaffected by rough conditions and electromagnetic disturbances, it has some very attractive features:

- Precise information about fire location, size, and propagation.
- It is easy to install.
- It is easy to integrate within a fire management system.

In addition to fire detection, there is also a rising demand for fire surveillance in facilities like rail and service tunnels, cable routes and ducts, storage facilities and warehouses, conveyor belts, and many other special installations operating in hazardous conditions. Manufacturers, such as LIOS Technology, offer systems tailored for these industrial applications with a benefit of achieving highly accurate temperature measurements over great distances at short measuring times combined with the ease in installation and data acquisition.

There are many more applications of the FO sensing and the literature on the subject is very extensive, but this brief review can give the reader a sense of the potential this technology can offer.

2

Distributed Fiber Optic Sensing

2.1 Introduction

Consider a single optical fiber that is connected to a sensing instrument and assume that this fiber is attached to a transmission tower. If now the tower is exposed to severe wind or ice loading, and an instrument connected to the fiber is able to provide strain or temperature information along the entire fiber length with a specific spatial resolution (say 1 m) and a prescribed accuracy, such a measurement is termed distributed sensing. Since a fiber is being used for sensing, it is called fiber sensor. Because the measured value is varying along the length of the optical fiber and providing the information spatially dispersed, it is termed distributed sensing.

Measurement of the temperatures and strains has been done in the past using thermocouples or strain gauges. Figure 2.1 illustrates a situation where both types of measuring devices were installed. Figure 2.2 shows comparison of the measurements made by traditional and fiber optic sensors. It shows how well a distributed optical fiber sensor provides strain and temperature along the aircraft's tail as it ascends from ground into the air and how well it compares with conventional technology.

Figure 2.3 shows an example of how a cluster of conventional strain gauges can be replaced with a single optical fiber sensor to provide the same information.

Even this brief review leads to a conclusion that the fiber optic sensing has several advantages over the traditional measurement technologies. This topic is explored in more detail below.

2.2 Advantages of the Fiber Optic Technology

Clearly, the use of a single optical fiber for measurement offers significant advantages over installation of multiple traditional sensors. Using a single optical fiber to provide the data along its entire length (up to 120 km!) with

Distributed Fiber Sensing and Dynamic Rating of Power Cables, First Edition.
Sudhakar Cherukupalli and George J. Anders.
© 2020 by The Institute of Electrical and Electronics Engineers, Inc.
Published 2020 by John Wiley & Sons, Inc.

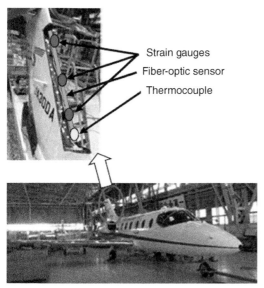

Strain gauges
Fiber-optic sensor
Thermocouple

MU-300 business jet used for the evaluation
and its mounted sensor portion

Figure 2.1 Photograph of a jet aircraft with optical fibers mounted in its wing and tail.

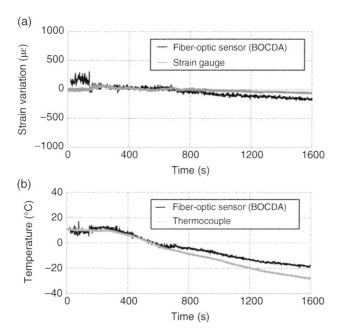

Figure 2.2 Comparative response of strain and temperature measured using conventional and optical sensors. (a) Strain variation. (b) Temperature variation.

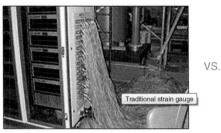

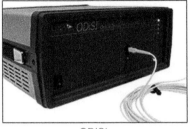

VS.

Traditional strain gauge   ODISI

Figure 2.3 Photograph shows the difference between multiple strain gauges to measure strain compared to an optical distributed sensing system.

strain and temperature information has the following advantages over the traditional methods:

- No need for multichannel data acquisition systems.
- Less clutter in the wiring is required as is obvious in Figure 2.3.
- There is no need to identify each channel and the placement of the sensors; thus, reducing errors in identifying the measurement points.
- An ability to multiplex data.
- A single fiber is presently capable of producing two vital pieces of information, i.e. temperature and strain, whereas with the conventional technology, one would need to deploy thermocouples, thermistors, or resistance temperature detectors (RTDs) and strain gauges.
- Calibration times will be much shorter compared to conventional technology.
- In case of a failure of the protected equipment, the damaged fiber will allow faster location of the failure point.
- It is immune to electromagnetic interference.
- Conventional thermocouples are generally prone to electromagnetic induction issues and, in such a case, shielded twisted pair thermocouples will have to be used.
- Conventional platinum RTDs are generally fragile and can get easily damaged by the electromagnetic forces. This is a common theme when these are used in a close proximity to the ac current-carrying cable.

However, there are also some disadvantages of the fiber sensing technology. These are discussed in Section 2.3.

2.3 Disadvantages of the Distributed Sensing Technology

One of the most significant disadvantages of the distributed sensing technology is the cost. If there is no need to measure temperature at several different locations, the deployment of conventional thermocouples (RTDs and thermistors) will be a fraction of the cost as compared to a FO system. In addition:

- The cost of a multichannel strain gauge monitoring system can be a fraction of a newer optical distributed strain monitoring system.
- Optical fiber cable systems are more fragile and prone to breaks if not properly handled and deployed.
- Limited temperature range if standard communication fiber is used.
- Time necessary to compute and display temperature results when fiber is of significant length.
- Time necessary to compute and display results when higher temperature resolution is required.
- Time necessary to compute results when lower spatial resolution is required.

Despite significant costs, the use of the distributed fiber sensing technology is growing quite rapidly. Its areas of application are discussed next.

2.4 Power Cable Applications

One of the main reasons why power cable temperature measurement of is desirable is because it very strongly influences the transmission capacity as the cable ampacity is primarily determined by the allowable conductor temperature. More accurate determination of this temperature can lead to better asset utilization, or conversely, reduced risk of overheating and related loss of life. Conductor temperatures are traditionally difficult to measure because it operates at a very high voltage and is surrounded by electrical insulation. In addition, power cables are most often either buried beneath the ground or under water and, therefore, are relatively inaccessible. The measurements are complicated because power transmission cables are often installed over long distances, passing through many different thermal environments with varying burial depths, thermal resistivities, and ambient temperatures. DTS systems overcome these challenges and provide cable temperatures along the entire cable corridor in real-time with good accuracy and spatial resolution.

One of the most relevant advantages of the optical fiber compared to the traditional wired temperature sensors is the absence of electromagnetic coupling to parallel power cables. Traditional temperature measurements using a long string of discrete sensors such as thermistors, thermocouples, and RTDs could be subject to interference, errors or malfunctions, and damage could occur because of the electromagnetic transients radiated from the cable system.

It is noteworthy that DTS systems have been used for power cable temperature profiling in North America since 1992 (see discussion in Chapter 1). Since then, important technical improvements have been made, for example, the 1992 applications were limited to monitoring only 2 km of the cable. Many power utilities and cable manufacturers have now realized the benefits recognized early by some utility engineers (CIGRE 2015; Hennuy et al. 2014; IEEE Standards 2012). Most modern DTS systems now include inboard PCs and have advantages of better environmental performance, higher accuracy, longer ranges, faster processing times, improved user interface and programming software, as well as capabilities to integrate with the dynamic rating systems (DRS). They can also allow control center operators to load cable circuits optimally using a variety of communication protocols.

In addition to temperature monitoring, DTS systems are increasingly used to monitor mechanical vibrations and actions that may lead to a physical damage to the buried power cable.

There are now many mainstream suppliers of DTS systems, while other smaller companies are breaking into the market, particularly for higher volume applications such as fire protection sensing, downhole sensing for the oil and gas industry, and so on. Table 2.1 provides a contact list of mainstream suppliers appropriate for the high-voltage power cable industry, mainly taken from their websites and promotional literature. A healthy skepticism is justified when reviewing some of the suppliers' claims, as there is a delicate balance to be struck between the cost, temperature and spatial resolution, range, and calculation times.

The large variety of DTS suppliers provides good competition, but it also presents difficulties in preparing project specifications, when there are many differences between them in terms of costs, technical performance, accuracy, and so on, while still aiming toward somewhat standardized applications at a particular utility. With an effort toward standardization in mind, there is a need to develop a coherent technical specification for a DTS system. based on many of the topics described herein.

Table 2.1 Contacts for mainstream DTS system suppliers.

Company name and source	Contact	Web address
AP Sensing, Germany	info@apsensing.com	www.apsensing.com
LIOS Technology, Germany	info@lios-tech.com	www.lios-tech.com/US
Omnisens, Switzerland	sales.us@omnisens.com	www.omnisens.com
Sensa (Schlumberger), U.K.	gparker2@slb.com	www.sensa.org
Sensornet, U.K.	enquries@sensornet.co.uk	www.sensornet.co.uk
Sensortran (Haliburton), U.S.	sales@sensortran.com	www.sensortran.com
Sumitomo Electric Industries, Japan	opthermo@jpowers.co.jp	www.sumitomoelectricusa.com
Roctest, Canada	—	www.roctest-group.com

Procurement, installation, and testing of DTS systems are certainly important and can be relatively complex, but at least equally important is deployment of the sensing fibers to obtain optimum results. Since the majority of new high-voltage cables are equipped with fiber optic sensors, the question arises how to employ FOSs in the existing lines.

The main opportunity for retrofitting existing cable systems is where the power cables are installed in ducts and there is a vacant or underused duct free for deploying an optical fiber cable. Although maximum benefit is obtained when the sensing fiber is as close to the energized cable core as possible, very useful information can still be obtained if the fibers are installed on the cable surface or in an adjacent vacant duct. This topic is discussed in more detail in Chapters 4 and 5.

It is important to remember that although DTS systems are capable of providing temperature profiles along an entire cable route, there are typically only several locations along the routes that are thermally limiting at maximum loadings and maximum ambient temperatures. DTS software provides the ability to create "zones" of interest along the various cable corridors from end-to-end, for monitoring and alarming. Zones can also be created to monitor ambient soil and air temperatures, remote from the power cables, which is valuable if implementing a DRS. For those applications with permanently installed DTS systems and SCADA connections, specific locations that are anticipated to exhibit higher zone temperatures can be preselected and the maximum zone temperatures reported to the system control center, along with the ambient temperature zone data.

The next evolutionary step is to use the zone temperature information to calculate the magnitude of power that can be transmitted in these corridors safely without excessively heating the cable and violating the zone temperature limit. In this way, the DTS-based DRS can be provided, with the dynamic ratings calculated either on board the DTS unit, on an adjacent computer, or at a control center. Implementation details for DRS will be discussed in later chapters of this book.

3

Distributed Fiber Optic Temperature Sensing

This chapter reviews the basic principles of the fiber optic temperature sensing. It is not meant to exhaustively analyze the physics of the DTS measurements but rather to give an understanding of the principles and applicability of various measurement methods used in practice.

3.1 Fundamental Physics of DTS Measurements

DTS systems inject a narrow laser pulse into an optical fiber through a directional coupler. A small amount of light is reflected back to the source (backscattered) as the pulse propagates through the fiber because of changes in glass density and composition as well as molecular and bulk vibrations caused by the heat. In a homogeneous fiber, the intensity of the backscattered light decays exponentially with time. Since the velocity of light propagation in the optic fiber is well known, the local position of the temperature is determined by measuring the arrival time of the returning light pulse (so-called time-of-flight) similar to a radar echo showing the distance of a car or plane. The concept is illustrated in Figure 3.1.

The returning backscattered light consists of different frequency spectrum components because of varying interaction mechanisms between the propagating light pulse and the optical fiber. These backscattered spectral components include Rayleigh, Brillouin, and Raman peaks or bands as shown in Figure 3.2.

3.1.1 Rayleigh Scattering

The Rayleigh backscattered component is the strongest because of density and composition fluctuations and has the same wavelength as the primary laser pulse. The Rayleigh component controls the main slope of the intensity decay curve and may be used to identify breaks and heterogeneities along the fiber. However, it is not sensitive to temperature.

Distributed Fiber Sensing and Dynamic Rating of Power Cables, First Edition.
Sudhakar Cherukupalli and George J. Anders.
© 2020 by The Institute of Electrical and Electronics Engineers, Inc.
Published 2020 by John Wiley & Sons, Inc.

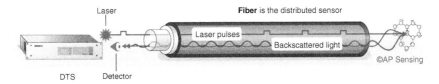

Figure 3.1 Determining the position of the analyzed signal. (*Source:* with permission from AP Sensing.)

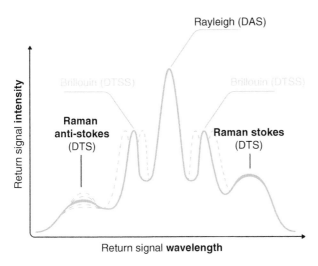

Figure 3.2 Backscattered spectral components.

3.1.2 Raman Spectroscopy

The Raman backscattered components are caused by thermally influenced molecular vibrations from the incident propagating light pulse and their intensity depends on temperature. Raman backscattered light has two components that are symmetric to the Rayleigh peak: the Stokes and anti-Stokes peaks. The intensity of the anti-Stokes peak is lower than that of the Stokes peak, but is strongly related to temperature, whereas the intensity of the Stokes peak is only weakly related to temperature. By calculating the ratio of the anti-Stokes to Stokes signal intensities, an accurate temperature measurement can be obtained. Combining this temperature measurement technique with the distance measurement through time-of-flight, the DTS provides temperature measurements incrementally along the entire length of the fiber.

3.1.3 Brillouin Scattering

The Brillouin backscattered components are caused by lattice vibrations from the propagating light pulse. However, these peaks are spectrally so close to the primary laser pulse that it is difficult to separate them from the Rayleigh signal. Because of this issue, systems using Brillouin technology are more complex and expensive, use only single-mode fibers, and, generally, have less accurate spatial and temperature resolutions than those based on the Raman backscattered light. Adopters of the Brillouin technology-based DTS systems claim to be capable of providing temperature profiles for open-ended fiber distances of 100–200 km with the aid of optical amplifiers (repeaters). Figure 3.3 illustrates how changes in the Brillouin frequency can be applied by some DTS systems to estimate fiber temperature as well as mechanical strain.

A principle of detecting mechanical vibrations is shown in Figure 3.4.

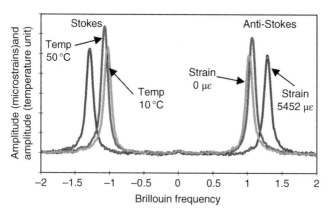

Figure 3.3 Changes in Brillouin frequency with temperature and strain (red and green Stokes lines show frequency shift of Brillouin light corresponding to temperature; and green and blue anti-Stokes lines show frequency shift corresponding to strain).

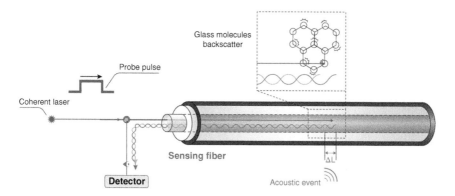

Figure 3.4 Detecting mechanical vibrations. (*Source:* with permission from AP Sensing.)

Figure 3.5 A typical Brillouin based DTS system (*Source:* Courtesy – LIOS Technologies.)

Figure 3.5 shows a picture of a Brillouin-DTS interrogator device and its basic detection scheme. A distributed feedback (DFB) laser generates narrow bandwidth radiation at 1550 nm that is amplitude-modulated by an optical modulator and amplified by an optical amplifier, in order to achieve the optimum pulse shape and power level for detecting spontaneous Brillouin scattering in a single-ended optical fiber. The scattered light is directed toward the detection system by an optical circulator. The weak Brillouin-scattered light is split into its polarization components and each component is mixed with a correspondingly polarized portion of DFB laser radiation serving as a local oscillator in a heterodyne detection scheme.

This polarization-diversified detection scheme measures all scattered light and is thus clearly more sensitive than a single-polarization detector. The weak Brillouin signal is mixed with a local oscillator, which is generated by the DFB laser light source, and detected by two photo receivers in an optical heterodyne scheme. Coherent detection of spontaneous Brillouin scattering is a proven measurement principle that ensures maximum detection sensitivity. The digitized raw data is averaged by a field programmable gate array (FPGA) and subsequently processed by an embedded computer. Measurement, averaging, and temperature calculations are run in parallel.

Therefore, the system continuously measures the Brillouin scattering without losing time for calculations.

3.1.4 Time and Frequency Domain Reflectometry

DTS systems typically have the following five major hardware components:

- Optical fiber.
- Laser.
- Photodetector and optical-electrical processing unit.
- Controller.
- Optical multiplexing switch (sometimes external) allowing temperature profiling of more than one fiber optic cable, at almost the same time.

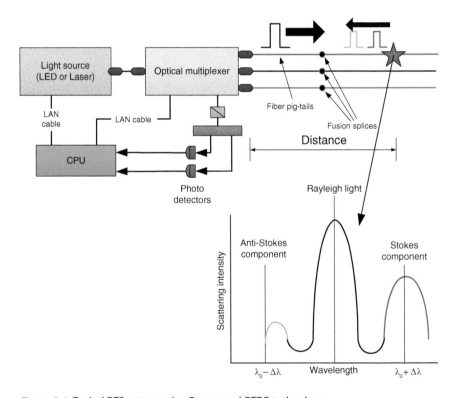

Figure 3.6 Typical DTS system using Raman and OTDR technology.

In addition, the software integrating all the above components and providing flexible user and SCADA interface is very important. It varies considerably between different DTS system providers.

A schematic description of a typical DTS system using Raman and optical time domain reflectometry (OTDR) technology is shown in Figure 3.6.

One supplier of the DTS equipment is using optical frequency domain reflectometry (OFDR) methods with Fourier transforms instead of the more traditional OTDR approach. They claim better accuracy, higher resolution, and less power consumption.

4

Optical Fibers, Connectors, and Cables

This chapter introduces the types of fibers available in the industry for DTS applications, the types of optical connectors, and their relevance in the context of distributed sensing applications. It also discusses how they may be incorporated into power cable transmission land and submarine corridors as well as the potential challenges that one may encounter during installation, particularly when fibers are embedded in the power cable itself.

4.1 Optical Fibers

Telecommunication has seen a significant change in the density, bandwidth, and speed which is essentially harnessed by light transmission through very fine glass or plastic fibers. Fiber optics has many advantages over copper wire, a comparison of some of the features is offered in Table 4.1.

The high signal bandwidth of optical fibers provides significantly greater information-carrying capacity. Typical bandwidths for multimode (MM) fibers range between 200 and 600 MHz-km and >10 GHz-km for single mode (SM) fibers. In contrast, typical bandwidths for electrical conductors are 10–25 MHz-km. Optical fibers are immune to electromagnetic interference and emit no radiation; they also are characterized by lower cost, size, and weight. Compared to copper conductors of equivalent signal-carrying capacity, fiber optic cables are easier to install, require less duct space, weigh 10–15 times less, and cost less than copper. Optical fibers have lower attenuation (loss of signal intensity) than copper conductors, allowing longer cable runs and fewer repeaters. Fiber optics do not emit sparks or cause short circuits, which is important in explosive gas or flammable environments. Since fiber optic systems do not emit RF signals, they are difficult to tap into without being detected. Fiber optic cables do not have any metal conductors; consequently, they do not pose the shock hazards inherent in copper cables. Optical fibers packaged in a cable form with

Distributed Fiber Sensing and Dynamic Rating of Power Cables, First Edition.
Sudhakar Cherukupalli and George J. Anders.
© 2020 by The Institute of Electrical and Electronics Engineers, Inc.
Published 2020 by John Wiley & Sons, Inc.

Table 4.1 Comparative performance of copper wire and fiber optic cables.

	Coaxial cable	Multimode fiber	Single-mode fiber
Product bandwidth distance	100 MHz-km	500 MHz-km	1 GHz-km
Attenuation (dB/km) at 1 GHz	>45	>1 dB	0.2 dB
Cable cost/km	—	$	$
Cable diameter	25 mm	3 mm	3 mm
Data security	Poor	High	High
Electromagnetic immunity	Poor	High	High

all dielectric materials (commonly referred to as all dielectric self-supporting [ADSS] cables) are being increasingly used in transmission corridors to meet telecommunication needs.

4.1.1 Construction of the Fiber Optic Cable and Light Propagation Principles

A single optical fiber is made of several concentric layers, as described in Figures 4.1–4.3, namely:

- Core: This central section, made of silica or doped silica, is the light-transmitting region of the fiber.
- Cladding: This is the first layer around the core. It is also made of silica, but not the same composition as the core. This creates an optical waveguide which confines the light in the core by total internal reflection at the core–cladding interface.
- Coating: The coating is the first nonoptical layer around the cladding. The coating typically consists of one or more layers of polymer that protect the silica structure against physical or environmental damage. The coating is stripped off when the fiber is connected or fusion spliced.
- Buffer: This buffer is typically 900 μm thick and protects the fiber from breaking during installation and termination and is located outside of the coating.

A coherent light source (LS) that injects light into the optical fiber is "guided" down the core of the fiber by the optical "cladding" which has a lower refractive index (the ratio of the velocity of light in a vacuum to its velocity in a specified medium) that traps light in the core resulting from "total internal reflection." This is schematically illustrated in Figure 4.4.

In fiber optic communications, SM and MM fiber constructions are used depending on the application. In MM fibers, the light travels through the fiber following different light paths called "modes," whereas in a SM fiber, only one

Typical layers of an optical fiber

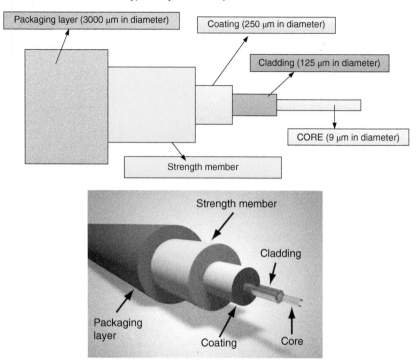

Figure 4.1 Schematic of a single-mode fiber and photo of the layers of an optical fiber.

Size comparison of a SM fiber

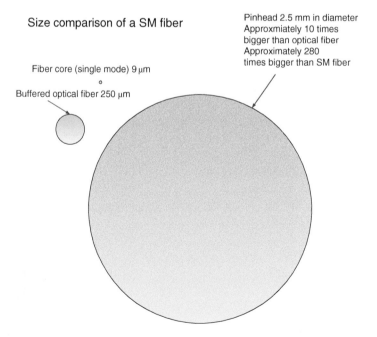

Figure 4.2 Size comparison of an optical fiber.

Figure 4.3 Photograph showing various fiber cable types (loose tube/tight buffer) (*Source: from BC Hydro*).

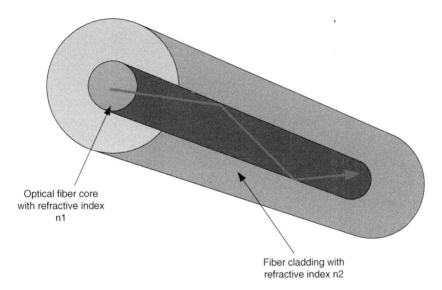

Optical fiber core
with refractive index
n1

Fiber cladding with
refractive index n2

Figure 4.4 Diagram showing total internal reflection resulting from the adjustment of the refractive indices of the two materials, namely the glass and the cladding.

mode is propagated "straight" through the fiber as illustrated in Figure 4.5a and b, respectively.

When light travels through the optical fiber, it is gradually attenuated with distance and this attenuation value is expressed in dB/km. Attenuation is a function of the wavelength (λ) of the light.

(a)   (b)

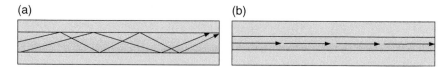

Figure 4.5 Diagram showing modes of optical transmission in (a) multimode and optical transmission in (b) single-mode fibers.

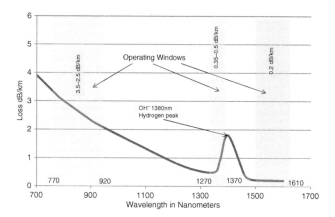

Figure 4.6 Attenuation vs. wavelength of optical fiber propagation (according to the graph).

Figure 4.6 shows the attenuation as a function of the wavelength. In this graph, the attenuation is 3 dB/km at 850 nm, reduces to 1 dB/km at 1310 nm, and reaches 0.5 dB/km at 1550 nm. There is a significant attenuation between 1310 and 1550 nm due to the hydrogen peak.

Bandwidth is a measure of the data-carrying capacity of an optical fiber. It is expressed as the product of frequency and distance. For example, a fiber with a bandwidth of 500 MHz-km can transmit data at a rate of 500 MHz along 1 km of fiber. The bandwidth of SM fibers is much higher than in MM fibers. The main reason for the lower bandwidth in MM fibers is modal dispersion.

As illustrated in Figure 4.7, in MM fibers, information (123) is propagated in according to N modes or paths as if it were "duplicated" N times (for example,

1 2 3 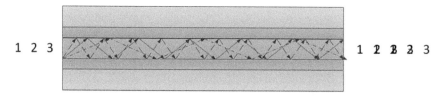 1 2 3 2 3

Figure 4.7 Modal dispersion in a fiber with fuzzy output.

in the diagram, the mode 3 path is longer than the mode 2 path, which are both longer than the mode 1 path). If data is too close, there is a risk of overlapping ("smearing") the information, and then it will not be recoverable at the end of the fiber. It is necessary to space the data sufficiently to avoid overlap, that is, to limit the bandwidth. Figure 4.7 illustrates diagrammatically what happens when digits 1, 2, and 3 are introduced into MM fiber (say red, yellow, and green laser beams); they become blurred at the receiving end due to modal dispersion.

In telecommunications today, typical operating wavelengths are 850 and 1300 nm in multimode, and 1300 or 1550 nm in a single mode. Note that there are natural "dips" in the attenuation graph at these wavelengths. For example, at an 850 nm operating wavelength, there is 3 dB attenuation after 1 km propagation (according to the graph). Three decibel of attenuation means that 50% of light has been lost (see Figure 4.6).

In a step-index fiber, the index of refraction changes abruptly from the core to the cladding. To help reduce modal dispersion, fiber manufacturers created graded-index fiber. This fiber has an index of refraction which gradually increases as it progresses to the center of the core. Light travels slower as the index of refraction increases. Thus, a light path propagating directly down the center of the fiber has the shortest path but will arrive at the receiver at the same time as light that took a longer path due to the graded-index of the fiber. This is illustrated in Figure 4.8.

Of course, modal dispersion is not an issue in SM fiber because only a single mode is propagated (see Figure 4.9).

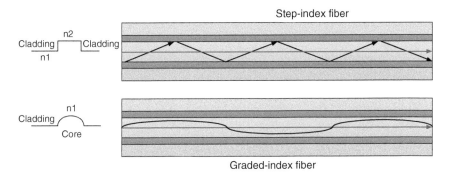

Figure 4.8 Step-index and graded-index fiber to minimize modal dispersion.

Figure 4.9 Single-mode propagation.

Unfortunately, the optical fiber construction shown earlier in Figure 4.3 is fragile. Thus, for most applications, the fiber must be made into a cable. There are many ways to construct a cable (tight buffer, loose tube, gel filled, distribution, breakout, etc.). However, in our single fiber cable example (see Figure 4.10), the 250 μm coating is jacketed with a 900 μm buffer and built into a 3.0 mm outer sheath cable with aramid yarn (Kevlar™) as a strength member.

As a typical example, Figure 4.11 portrays a cable with multiple optical fibers.

4.1.2 Protection and Placement of Optical Fibers in Power Cable Installations

There are many different types of optical fibers that are manufactured for various applications. Some of the commonly used are:

- MM optical fibers (50 or 62.5/125/250 μm)
- SM optical fibers (9/125/250 μm)

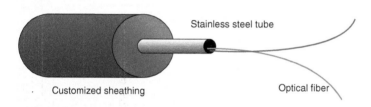

Figure 4.10 Protection of an optical fiber.

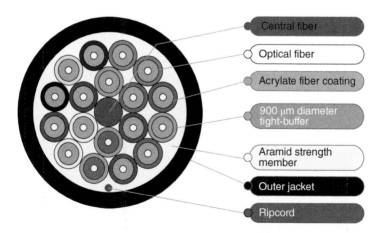

Figure 4.11 Bundled fiber cable with many different types of optical fibers.

- Polarization-maintaining fibers
- Dispersion-shifted fibers
- Hydrogen darkening-resistant fibers

Typically, for DTS applications either SM or MM fibers are most commonly used. SM fibers are used where temperatures have to be measured over long distances (to a maximum of about 40 km using single-mode Raman-based system, 70 km for Brillouin-based DTS system). MM fibers can be used when the measuring distances are less than 30–40 km.

Fibers can be placed in loose or tight buffer tubes (that are color coded), packaged mechanically supported, and protected in an all-dielectric construction referred to as ADSS cables. Such cable construction is preferred when placed in the proximity of high-voltage cables to avoid induction issues if power cables are armored with a metal sheath for mechanical protection. Fibers may also be placed in a metal tube that is typically 1.4–2 mm in diameter, made of stainless steel or copper where the measuring fibers might encounter high heat such as in downhole well monitoring or require mechanical protection such as when embedded in a power cable. Such constructions are referred to as fiber in metal tube (FIMT).

Some power cable manufacturers can provide optical fibers in metal tubes for integration into power cables during manufacturing. Typically, the FIMT is applied either under continuous metallic moisture barriers (sheaths) or between the metallic moisture barrier and anticorrosion jacket. For cables with a metal laminate moisture barrier with additional concentric wires to handle fault current, the FIMT can replace several concentric wires. For submarine cables, they can also replace several armor wires or embedded in the anticorrosion sheath (see Figure 4.12). In both cases, careful consideration should be given to the sheath design and sheath bonding protection.

Table 4.2 summarizes various options for placement of the optical sensing fibers for distributed temperature measurements (see also Section 5.1 for more details about this topic) and discusses some of the advantages and disadvantages of each.

The present industry consensus is that alternatives 2, 4, 5, 6, and 7 are the most practical. Many cable manufacturers are now providing cables with integrated FIMT sensing fibers as described in 4, 5, and 6, which are recommended for future applications, provided initial costs are not prohibitive. A variation for submarine cables is replacing several armor wires with FIMT tubes containing sensing fibers.

A critical requirement is that the fibers within the tubes, as well as the tubes themselves, are capable of withstanding tensions, bending and sidewall pressures applied during manufacturing, transportation, and installation, and a lifetime of operation. During operation, load cycling from ambient to maximum temperatures can occur. Therefore, the design of the optical fiber

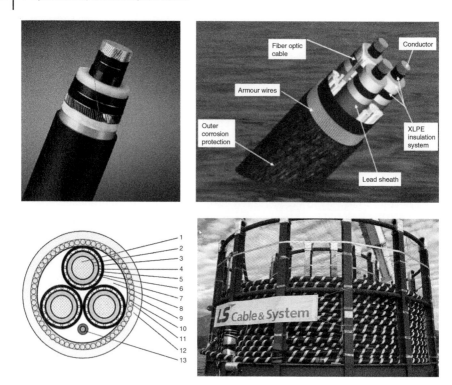

Figure 4.12 Optical fibers replacing concentric wire neutral for a single conductor power cable and located below armor bedding in a three-core submarine cable. (*Source:* courtesy of LS Cable and System.)

should be supported by test results. This typically requires that the FIMT has a "loose tube" design and that it takes a sinusoidal path along the cable axis, except where concentric metal shield wires or submarine cable armor wires are replaced. For these applications, the FIMT must follow the helical angle of the wires or armor.

Quite often, to ensure water does not ingress longitudinally into the cable, a gel or polymer is introduced in the loose tube.

Similarly, cable manufacturers must be able to confirm by tests that the presence of FIMTs is not detrimental to the long-term performance of the power cables.

In some cases, it may be necessary to provide a "guard" wire on either side of the FIMT, of slightly bigger diameter, to ensure protection from excessive sidewall crushing forces as illustrated in Figure 4.13.

Even though the majority of FO sensors installed in power cables are protected by the stainless steel tubes, sometimes copper tubes are used. Typically, they are made by continuous longitudinal seam welding of metal tapes around

Table 4.2 Options for placement of FO cables for retrofits and new applications.

Alternative no.	FO installation location	Advantages	Disadvantages
1	Retrofit fiber in metal tube (FIMT) cables into a hollow conductor	Very accurate conductor temperature measurement – the ultimate goal	Insertion can only be done at the terminations. Termination modifications needed. Special FIMT optical fiber cable is needed. Insulating downlink bushings needed to bring the sensing fibers to the ground potential. Difficult to insert fiber for more than about 100 m, which probably is insufficient for locating thermally limiting sections of the underground circuits
2	Retrofit sensing fibers into a vacant duct adjacent to power cables in a duct bank	Installation is relatively easy. Standard ADSS optical fiber cable can be used. Entire cable route can be covered, so that thermally limiting sections can be identified. Mitigation efforts can be aimed at correcting thermally limiting sections	A longer delay in detecting temperature changes because of load fluctuations compared to the case when the sensing fiber was inside or on the surface of the power cable
3	Incorporate FIMT cables into the cable conductor	Very accurate conductor temperature measurement – the ultimate goal. Very fast temperature response to changing load – advantageous for DRS	Risk of FIMT cable being damaged during cable manufacturing, because of a mechanical stress and high temperatures during the vulcanization process (XLPE). Power cable field joints and terminations are more complex because of a need to also splice the sensing fibers within the FIMT cables. Risk of damage to sensing fibers during field jointing and terminating
4	Incorporate FIMT cables under power cable sheath	More accurate conductor temperature inference. Fast temperature response to changing load – advantageous for DRS	Risk of FIMT fibers being damaged during cable manufacturing, because of high temperatures during sheath extrusion, longitudinal seam welding, or application of hot melt adhesives for laminated moisture barrier. Risk of damage or deformation of the underlying insulation, perhaps because of long-term thermomechanical effects at high operating temperatures. Risk of damage to sensing fibers during field jointing and terminating. FIMT needs to follow a sinusoidal pattern down the cable axis in order to accommodate strains during cable bending, manufacturing, transportation, installation, and operation for the circuit life

(Continued)

Table 4.2 (Continued)

Alternative no.	FO installation location	Advantages	Disadvantages
5	Incorporate FIMT cables in supplemental metallic shielding wires in cables with laminated moisture barriers	More accurate conductor temperature inference. Fast temperature response to changing load – advantageous for DRS	Risk of fibers being damaged during cable manufacturing, because of high temperatures during application of the jacket. Risk of damage to sensing fibers during field jointing and terminating. Helix length of supplemental shielding wires may not be sufficient to accommodate strains during cable bending, manufacturing, transportation, installation, and operation for the cable life. Positional information of hot spot locations requires careful review
6	Incorporate FIMT cables over power cable sheath and under jacket	More accurate conductor temperature inference. Fast temperature response to changing load – advantageous for DRS. Easier access to fibers than previous alternatives	Risk of fibers being damaged during cable manufacturing, because of high temperatures during jacket extrusion. Risk of damage to sensing fibers during field jointing and terminating. FIMT cable needs to follow a sinusoidal pattern down the cable axis in order to accommodate strains during cable bending, manufacturing, transportation, installation, and operation for the cable life
7	Fasten sensing cables onto surface of power cables during installation	Less accurate conductor temperature inference. Slower response to changing load – compromised value for DRS. Much easier access to fibers than previous alternatives. Easier power cable field jointing and terminating than previous alternatives. Easy to redirect sensing fiber cables to form ground and air loops to also determine ambient temperatures at key points along the route. Valuable for submarine cable shore ends, which are often thermally limiting	Mediocre conductor temperature inference. Less value for DRS because of slow temperature response to the changing load. There is a risk of false temperature reading if the fiber is not lying flat on the surface of the cable. In this case, the temperature reading may be lower by as much as 5–10 °C and, if this is a hot spot, it can prevent its detection

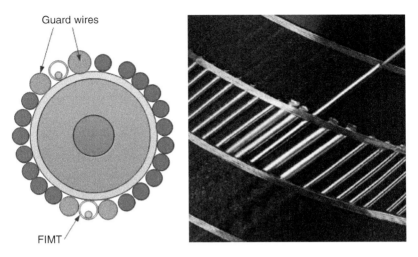

Guard wires

FIMT

Figure 4.13 Guard wires beside the optical fibers for mechanical protection replacing concentric wire neutral for aluminum and copper.

Sensor cable underground HDPE and FIMT (fiber in metal tube)

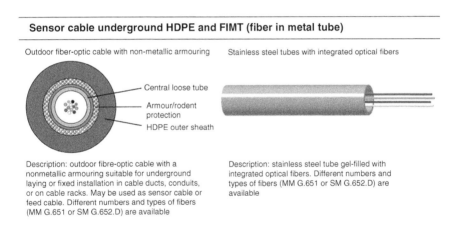

Outdoor fiber-optic cable with non-metallic armouring

Central loose tube

Armour/rodent protection

HDPE outer sheath

Stainless steel tubes with integrated optical fibers

Description: outdoor fibre-optic cable with a nonmetallic armouring suitable for underground laying or fixed installation in cable ducts, conduits, or on cable racks. May be used as sensor cable or feed cable. Different numbers and types of fibers (MM G.651 or SM G.652.D) are available

Description: stainless steel tube gel-filled with integrated optical fibers. Different numbers and types of fibers (MM G.651 or SM G.652.D) are available

Figure 4.14 Typical construction of optical fiber elements.

the fibers using laser methods, and the tube is filled with water-blocking compound and a hydrogen scavenging material. The longitudinal weld is checked for welding defects by means of eddy current technology, as used for the tubes encapsulating the optical fibers in trans-oceanic telecommunication submarine cables. In some cable construction, the optical fibers are also placed in a polyethylene tube. This tube could be oval in cross-section to minimize any increase in the power cable diameter.

Figure 4.14 and Table 4.3 provide typical specification of optical fibers.

Table 4.3 Typical specifications of optical fiber elements.

Typical values	Wavelength (nm)	Multimode G.651 (dB/km)	Single-mode G.652.D (dB/km)
Attenuation	850	≤2.4	
	1300	≤0.5	
	1310		0.35
	1383	≤0.5	
	1550		0.22
Bandwidth		MHz-km	
	850	≥440	
	1300	≥1000	
	1300	≥2000	

4.1.3 Comparison of Multiple and Single-Mode Fibers

MM fiber has a bigger core than the SM fiber and allows for more optical power to be launched into it, bringing higher temperature resolution. However, it can create more attenuation (due to higher backscattering), thus limiting the ability to monitor long distances. In addition, MM fiber suffers from modal dispersion which broadens the light pulse as it propagates. This is especially true over long distances when using the optical time domain reflectometry (OTDR)-based DTS system, thus affecting spatial resolution. SM fiber has a smaller core restricting the amount of optical power to be launched into it. However, backscattering losses are smaller ("one mode" of propagation of the light) allowing for the monitoring of longer distances.

Typically, DTS suppliers use lasers that operate at wavelengths ranging from 1064 to 1550 nm (outside of visible range – safety is a very important consideration during operation involving fiber splicing, terminating, and testing).

Some manufacturers suggest that distances greater than 30–40 km with MM fibers are possible but the user should verify the prevailing conditions to confirm that the spatial and temperature accuracy are indeed achieved. For longer distances, up to 70 km, SM fibers are employed. Industry standards are well developed. The standards are the same for both types of fibers with the exception of the standard listed in the bibliography substituting for ITU G.651.1.

It is very important that the fiber type chosen for field deployment (single or multimode) whether packaged as an ADSS or FIMT is compatible with the DTS equipment to be provided. It could be beneficial to standardize on a few fiber types and installation schemes (for example, choosing MM fiber for land cables or SM fiber for long submarine installations or ADSS or FIMT construction types), so that DTS equipment can be used flexibly at a variety of measurement sites. However, any one should use the best available instrument/ fiber type depending on the link architecture. For example, on land, the type of fiber/instrument is mostly determined by the length and power budget.

The measuring range is dependent on the following factors, but not limited to:

- Laser power in the DTS unit (mW).
- Optical loss (dB/km) along the fiber.
- The type of connector and connector losses (dB).
- Number and quality of splices along the fiber circuit.
- Attenuation per splice (dB).
- Fiber core size (μm).
- Wavelength of the laser (nm).

4.2   Optical Splicing

Fiber optic links require a method to connect the transmitter to the fiber optic cable and the fiber optic cable to the receiver. In general, there are two methods to link optical fibers, namely:

- Fusion splice
- Connector splicing

The fusion splice (FS) operation consists of directly linking two fibers by welding with an electric arc or a fusion splice and then mechanically protecting the assembly with a heat shrinkable jacket. The advantages of this approach are that the linking method is fast and simple and there is very little insertion loss (the loss of light generated by a connection is called insertion loss). The disadvantages are that the link is relatively fragile, is permanent, and the initial cost (of the fusion splicer) is high. Prior to the actual fusion operation, it is very important to cleave the fiber ideally at an angle (ranging between 8 and 10°) and then placing the two cleaved ends into a fusion splicer. Modern-day fusion splicers once activated will align the fibers in 3D injecting light into the fibers and ensuring good optical coupling before they are spliced and this operation can take barely a minute.

The second method involves the use of fiber optic connectors. A connector terminates the optical fiber inside a ceramic ferrule using epoxy to hold the fiber in place. The connectors can be mated and unmated at any time. The advantages of this approach are that the connection is robust, the connector can be chosen according to the application, and can be connected and disconnected hundreds or even thousands of time without damage. The disadvantages of this approach are that making a connection takes longer than fusion splicing, requires special tools, and the insertion loss can be higher when compared to fusion splicing. There are two types of fiber optic connectors: physical contact (PC) and expanded beam.

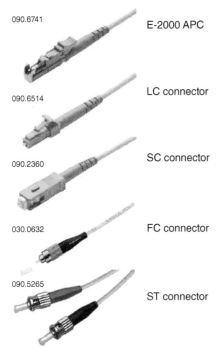

090.6741

E-2000 APC

090.6514

LC connector

090.2360

SC connector

030.0632

FC connector

090.5265

ST connector

Figure 4.15 Different types of connectors used in DTS Systems. (*Source:* courtesy R&M Systems.)

PC connectors utilize fiber in a precisely machined ferrule. This allows easy handling of the fiber and protects it from damage. The principle of PC connectors involves the direct contact of polished fibers within two ceramic ferrules. The ferrules are aligned using a ceramic alignment sleeve. Insertion loss is a function of the alignment accuracy and the polish quality. There are springs behind the ferrule to ensure that the two ferrules are in constant contact even in high vibration and shock environments.

There are many types of fiber optic connectors used in various applications (see Figure 4.15). The most popular are called ferrule connectors (FC). Although the FC are being replaced in many applications (telecom and datacom) by lucent connectors (LC) (see the definitions below) and subscriber connectors (SC), they are still used in the measurement equipment. The connector has a screw threading and is keyed allowing the ferrule to be angle polished providing low back reflection (light is reflected back to the transmitter, most often at the connector interface due to an index of refraction change).

LC are supplanting SC because of their smaller size and excellent panel packing density and push–pull design. They are also used extensively on small form-factor pluggable transceivers.

SC also offer a push–pull design (which reduces the possibility of the end-face damage when connecting) and provide good packing density. They are still used in datacom and telecom applications.

Straight tip connectors (ST) are engaged with a bayonet lock which is engaged by pushing and twisting the connector. The bayonet interlock maintains the spring-loaded force between the two fiber cores.

These PC connectors perform well against particle contaminants (dust, mud, and so on) and are usually less sensitive to liquid contaminants (water or oil). The PC pushes liquid out of the way and the liquid does not degrade the connection. PC connectors are cleaned by wiping the ferrule with a clean cloth or wipe, spraying with a cleaner, or washing with water.

4.3 Fiber Characterization

Traditionally, standard fiber optic cable testing used to be governed by the Telecommunication Industry Association (TIA) Optical Fiber System Test Procedure (OFSTP) has now been replaced by a new ISO standard. The TIA has adopted IEC 61280-4-1 as the replacement for OFSTP-14. The two documents are mostly the same, with some important exceptions. OTDR testing was also approved as a second-tier test method as long as both launch and receive cables are used. The reference test cables with "reference-grade connectors" were recommended. Methods are given for testing and verifying the loss of reference test cables.

Most of the changes are in the nomenclature. In the meantime, it has been recommended that the testing be continued as usual. There are five ways listed in various international standards from the EIA/TIA and ISO/IEC to test installed cable plants. Three of them use test sources and power meters to make the measurement, while the fourth and fifth use an OTDR.

The diagrams in Figures 4.16–4.19 provide an illustration of the various test methods. The source/power meter method, generally called "insertion loss," approximates the way the actual network uses the cable plant, so one would expect the loss to be similar to the actual loss seen by the network. The OTDR is an indirect method, using backscattered light to imply the loss in the cable plant, which can have large deviations from insertion loss tests. Optical time domain reflectometers (OTDRs) are more often used to verify splice loss or find damage to cables. The differences in the three insertion loss tests are in

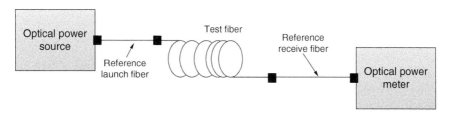

Figure 4.16 Typical power insertion loss measurements – Method 1.

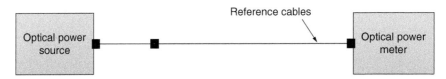

Figure 4.17 Typical power insertion loss measurements – Method 1 – alternative.

Figure 4.18 Typical power insertion loss measurements – Method 2. (The meter, which has a large area detector measures all the light coming out of the fiber, effectively has no loss and, therefore, measures the total light coming out of the launch reference cable.)

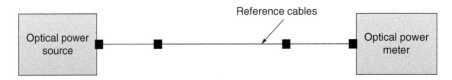

Figure 4.19 Typical power insertion loss measurements – Method 3.

how we define "0 dB" or no loss. All three tests end up with the same test set but the reference power can be set with one, two, or three cables as shown in the three setups in Figures 4.16–4.19. All four methods have measurement uncertainty.

Standard test methods use a LS and reference "launch" cable on one end of the "cable to test" and a "receive" reference cable connected to a power meter (PM) on the far end. The test is intended to measure the loss of the connections of the connectors on either end to the reference test cables and the loss of the rest of the cable (which may include splices or additional connections in addition to the fiber).

All standards offer three methods of setting a "0 dB" reference using one, two, or three reference cables. It is assumed that all reference cables are short enough (generally <2 m or 6 ft) that the loss of the fiber is low enough to be ignored and that the connectors are of high quality – that is to say low loss. It is also assumed that the PM has a large-enough detector that it gathers all the light from the end of a fiber so it has a consistent connection to the fiber being connected to its interface.

The three different methods evolved from three issues encountered in practice are:

1) The test equipment with certain connectors does not match the connectors applied on the test cable.
2) Technicians working at either end of a length of the test cable use different light sources which have no proper reference output (mW) which might serve as a reference.
3) Belief that since one needs a launch and receive reference cable, one should make the reference setting using both of them.

If one examines the schematic drawings of the three methods, it becomes evident that when the reference is set, one or two connections are included. This means that the loss of those connections is included in setting the 0 dB reference. This has led to the misunderstanding that the three cable reference, which includes two connections when the reference is set, does not measure the loss of the connectors at the ends of the cable because two connections were included in setting the 0 dB reference and the one cable reference (the old Method B) only measures one connector because one is included in setting the reference.

It is important to note that any version of the test measures all the connector losses in the cable under test, but subtracts the loss of connections included when setting the 0 dB reference.

Insertion loss testing with a test source and PM simulates the way the cable system will be used with an actual measurement link. The test source mimics the transmitter, the PM, and the receiver. But insertion loss testing requires reference cables attached to the source and meter sides to connect to the cable under test as shown in Figure 4.16. This insertion loss test can use one, two or three reference cables to set the "0 dB loss" reference for testing. Each way of setting the reference gives a different loss.

Generally, network standards prefer the one reference cable loss method, but it requires that the test equipment uses the same fiber optic connector types as the cables under test. If the cable has different connectors than the test equipment (e.g. LC-type connector on the cable and SC-type connectors on the tester), it may be necessary to use a two or three cable reference, which will give a lower loss since connector loss is included in the reference and will be subtracted from the total loss measurement. Any of the three methods are acceptable, as long as the method is documented.

In Figure 4.17 there is no reference fiber but only a reference launch fiber and the fiber to be tested.

The second method uses only one reference cable attached between the test source and the PM. The meter, which has a large area detector measuring all the light coming out of the fiber, effectively has no loss and, therefore, measures the total light coming out of the launch reference cable. Once the reference is set, the launch reference cable should not be removed from the source, as it may have a different coupled power when reattached.

When the cable is tested as shown above, the loss measured will include the loss of the reference cable connection to the fiber optic cable under test, the loss of the fiber and all the connections and splices in the cable plant, and the loss of the connection to the reference cable attached to the meter.

It is important to note that with all these tests the quality of the reference cables is important to avoid the uncertainty of the measurement. If one uses reference cables with bad connectors, the losses when mated to the cable under test will be higher than they should be, not a good result if one wants the installed cable to show the quality of the installation processes! Installers should test all reference cables using FOTP-171 single-ended cable tests to ensure the required quality. Cables with losses higher than 0.5 dB per end should be cleaned and retested, then discarded if they do not meet the 0.5 dB max loss. If any of the connectors are dirty, measurements will show higher loss and more variability.

All four methods have measurement uncertainty, which is examined after we explain the methodology.

The testing method probably evolved from the telecommunication companies where optical cable lengths were long and the instruments were brought together only once a day for calibration, plus instruments often had connectors different from the cable plant (e.g. biconics on the cable plant and SMAs on the instruments). Here, the launch reference cable is attached to the source, the receive reference cable to the meter, and then the two cables are mated to set the reference.

Setting the reference this way includes one connection loss (the mating of the two reference cables) in the reference value. When one separates the reference cables and attaches them to the cable under test, the decibel loss measured will be smaller by the connection loss included in the reference setting step. However, this is an approximation since the variations in mating alignment make the loss slightly different each time two connectors are mated.

Method 1 gives a smaller loss than the one cable reference and Method 2 described above has a higher variability because one connection is included in setting the reference. If any of the connectors are dirty, measurements will show higher loss and more variability. If the connectors are dirty when setting the reference but cleaned afterwards before testing the cable, one gets a lower loss, or even a gain!

If the test equipment has different connectors than the cable under test, but the connectors are all mateable with proper mating adapters, this method can be used.

Some of the newer connectors are male/female or plug/jack, not two males that use a mating adapter to create a connection. One connector is used on the jack in the wall or patch panel and one is used on a patch cord. These connectors cannot be mated to test equipment or can two similar ones (plug or jack) be mated with a mating adapter. Reference cables generally will be patch cords with plugs while the cable under test will have jacks on either end. The only way to get a valid reference is to use a short cable of known good condition as a "stand-in" for the cable to be tested to set the reference. To test a cable, the reference cable is replaced with the cable to test and a relative measurement is made.

Obviously, this method includes two connection losses in setting the reference, so the measured loss will be reduced by the two connection losses and has higher uncertainty. Once again, if any of the connectors are dirty, measurements will show higher loss and more variability. If the connectors are dirty when setting the reference but cleaned afterwards before testing the cable, the effects could be as stated above.

The need for such optical attenuation measurements stems from the fact that if the test equipment has different connectors than the cable under test and the connectors on the cable to test are not mateable, this method can be used. Since this method works with any connector style, it has been chosen for several international test standards.

Since there is some confusion about what each method actually measures, the FOA has a web page that explains it in simple diagrams and math for testing fiber optic cable plant.

Insertion loss testing with a test source and PM simulates the way the cable plant will be used with an actual link. The test source mimics the transmitter, the PM the receiver. But insertion loss testing requires reference cables attached to the source and meter to connect to the cable under test. This insertion loss test can use one, two or three reference cables to set the "zero dB loss" reference for testing. Each way of setting the reference gives a different loss.

OTDRs test from one end of the cable using the backscatter signature of the fiber to make an indirect measurement of the fiber (see Figure 4.20). OTDRs always require a launch cable for the instrument to settle down after reflections from the high-powered test pulse overloads the instrument. OTDRs have traditionally been used with long-distance networks where only a launch cable is used, but this method does not measure the loss of the connector on the far end. Adding a reference cable at the far end allows measuring the loss

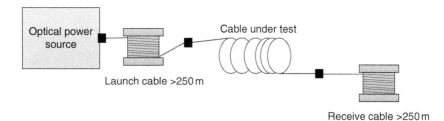

Figure 4.20 Typical setup for an OTDR test.

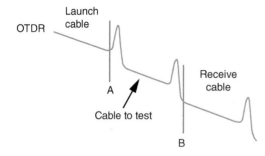

Figure 4.21 Typical OTDR trace showing the launch cable, cable under test as well as the receive cable and the connectors which can amplify or attenuate the light.

of the entire cable, but negates the big advantage of the OTDR, that it makes measurements from one end of the cable only, since a technician is required to attach the receive cable to each fiber as it is being tested.

When testing with an OTDR using only the launch cable, the trace will show the launch cable, the connection to the cable under test with a peak from the reflectance from the connection, the fiber in the cable under test, and likely a reflection from the far end, if it is terminated or cleaved (see Figure 4.21). Most terminations will show reflectance that helps identify the ends of the cable.

Using markers A (set at the end of the launch cable) and B (set at the end of the cable under test), the OTDR can measure the length of the cable under test and the loss of the connection to the cable under test plus the loss of the fiber in the cable and any other connections or splices in the cable under test.

This method does not, however, test the connector on the far end of the cable under test because it is not connected to another connector, and connection to a reference connector is necessary to make a connection loss measurement.

If a receive cable is used on the far end of the cable under test, the OTDR can measure the loss of both connectors on the cable as well as the fiber in the cable and any other connections or splices in the cable under test. The placement of the B marker after the connection to the receive cable means some of the fiber in the receive cable will be included in the loss measured. Most OTDRs have a "least squares" test method that can subtract out the cable included in the measurement of a single connector, but this will not work on a double-ended cable loss test.

Everything the OTDR learns about the fiber is dependent on the amount of light scattered back toward it and how the instrument is set up for the test. This "backscatter" is a function of the materials in the fiber and the diameter of the core. Joints between two dissimilar fibers that have different backscatter coefficients will not allow one-way measurements. One way the loss is too high, the other way too low (maybe even a "gainer" where the change in backscatter is more than the loss of the connection).

The second problem with OTDRs on MM fiber is the laser source. As mentioned above, lasers couple light narrowly into MM fiber and will measure lower attenuation and connector or splice loss than recommended by standards on the outward-bound test pulse, but scattered light probably overfills the fiber, even more than a LED on the return. To date, we are unaware of anyone who has modeled this and can provide guidance on the expected test results from an OTDR.

In addition, there are problems in premises applications with OTDR distance resolution. Light travels about 1 m in 5 ns. The width of the test pulse is usually 10–30 ns and the minimum resolution of the OTDR is about three times that or 2–6 m. Highly reflective events like multimode connectors in premises cabling cause instrument overload and lengthen the minimum resolution of the instrument. Only a few specialized OTDRs have the resolution needed for premises cabling.

It has to be noted that a DTS is manufactured and factory tested with a fiber cable available to the manufacturer. The on-site fiber optic sensor may have different parameters and characteristics. While it is not a problem for communication, it may provide the erroneous temperature reading. The equipment should be adjusted to the installed sensor during commissioning by a knowledgeable technician before a measurement is taken and the temperatures confirmed.

OTDRs are complicated instruments. Before the OTDR is used to make a measurement, all these parameters have to be set correctly: range, wavelength, pulse width, number of averages, index of refraction of the fiber, and the measurement method (usually two types for each measurement). OTDR manufacturers should teach the user how to set up the OTDR properly and how to interpret the rather complicated display. But few customers are willing to invest the day or two necessary to learn how to use the instrument properly. So manufacturers create an "autotest" function like a Cat 6 certifier that tests the fiber and gives a pass/fail result. Technicians need to be trained in OTDR testing to allow proper data interpretation.

Unfortunately, because of their indirect measurement technique, OTDRs do not easily correlate with insertion loss tests, and therefore they are not allowed by industry standards to be used alone. Some users claim to have been able to control modal power in MM fiber and get correlation between OTDRs and insertion loss tests, but results are hard to duplicate. OTDRs have been known to give divergent results between two different units!

If one considers the OTDR test to be a "qualitative" not "quantitative," and one knows how to interpret the OTDR trace properly, one can determine if connectors and splices are properly installed and if any damage has been done to the cable during installation. If the user does not have the experience and knowledge to do a proper analysis, the device usually only causes problems.

Table 4.4 shows test results from measurements done on a 520 m simulated test cable plant with MM 62.5/125 fiber tested four ways at 850 nm using

Table 4.4 Comparative loss on a test cable length using different methods.

Test method	Loss (dB)	Standard deviation (dB)
Reference cable 1	2.96	±0.02
Reference cable 2	2.66	±0.02
Reference cable 3	2.48	±0.024
With launch reference only	1.91/2.05 (in reverse direction)	

several different sets of reference cables. Using 10 different sets of reference cables and making multiple measurements allowed calculating an average of each measurement, comparable to what several different test crews might find, plus calculating the standard deviation, a statistical indication of the repeatability of the measurement.

Clearly, the loss of the cable under test reflects the comments we made above. The first reference method has higher loss than the other methods, but it also has much less measurement uncertainty (standard deviation.) The two and three cable reference methods have less loss because we have subtracted the connector loss(es) included when we set the reference for 0 dB loss, and the uncertainty is higher because of the greater variance when connected to the reference cables. The OTDR measurement is significantly lower than the other three methods plus is 0.14 dB different depending on in which direction we are testing. Note that if we added to the OTDR loss the average connector loss difference in the other tests, about 0.25 dB, the loss would still only be about 2.1–2.2 dB, significantly smaller than any of the insertion loss tests.

All four measurement methods are performed strictly according to the international standards, so the dilemma is which test result or method should be adopted to ascertain the optical loss and there is no clear choice!

Most network specifications have been written around a test that uses the one cable reference approach, as does the EIA/TIA 568B test method. If the cable has been tested with any other method, the lower loss that could be measured will give a false indication of the system margin. If we are dealing with new, fast networks like Gigabit Ethernet or Fiber Channel, which have much smaller operating power margins, this can spell trouble.

The significantly different optical attenuation (loss) obtained in OTDR measurements comes from its completely different methodology. OTDRs use backscattered light to imply the measurement while insertion loss is measured directly by transmission. All OTDRs use laser sources which inject light generally only in the center of the core of the fiber where the loss is lower. Some claims for using mode conditioners to make OTDR and insertion loss data correlate could not be proven by independent tests conducted by the Fiber Optics Association. Recent standards mention and sometimes require a receive cable for OTDR tests, which would have added 0.2–0.3 dB to the loss measured in the test detailed above. But having to use a receive cable with an OTDR negates its advantage of being able to be tested from one end by a single technician.

The results of the insertion loss testing will depend on the types of connectors on the test equipment and the connectors on the cables being tested. If the tested cable has the same connector styles as the test set, or all the connectors use the same 2.5 mm ferrule style common to ST, SC, FC, and some other connectors, the one cable reference method should be used.

On the other hand, if the test equipment has ST connectors or SCs and one must examine LCs on the cable under test, one may have no choice but to use a two cable reference or three cable reference approach. To compare these test results with a one cable reference, one needs to add an estimated loss for the connector included in the reference measurement, say 0.3–0.5 dB for a typical connector on a high-quality factory-made patch cord.

It has to be noted that a DTS is manufactured and factory tested with a fiber cable available to the manufacturer. The on-site fiber optic sensor may have different parameters and characteristics. While it is not a problem for communication but it may provide erroneous temperature readings. The equipment should be adjusted to the sensor during commissioning by a knowledgeable technician before a measurement is taken and the temperatures confirmed.

4.4 Standards for Fiber Testing

TIA is accredited by the American National Standards Institute (ANSI) as a "standards developing organization (SDO)." Standards projects and technical documents initiated by TIA's engineering committees are formulated according to the guidelines established by the ANSI Essential Requirements and the following documents:

- PANS document (PDF): contains comprehensive TIA procedures for developing American National Standards.
- TIA Engineering Committee Operating Procedures (PDF): contains comprehensive operating procedures for TIA Engineering Committee Meetings.
- Guidelines to the Intellectual Property Rights Policy of TIA (PDF): a companion document to the Engineering Manual.
- TIA Committee Style Manual (PDF): contains instructions for styling TIA standards documents that are in development (e.g. figures, tables, and indexes).

Every new standard project begins as a technical contribution in one of TIA's technology-oriented engineering committees or subcommittees. A Project Initiation Notice (PIN) form is completed and submitted for approval by TIA staff. After the project is approved for initiation, the formulating engineering committee or subcommittee will continue to develop the technical parameters of the project. When the proposed standard nears completion, the engineering group will chose to submit the document for approval as a "TIA Standard," an "American National Standard," or both.

Standards in the optical fiber community are constantly updated because of the new knowledge from research activities, Federal Communication Regulations, and new technological developments. A full catalog of TIA's standards may be obtained by visiting the website: www.tiaonline.org.

4.4.1   Fiber Optic Testing

The following is a listing of different standards for optical fiber testing.

FOA Standard FOA-1: Testing Loss of Installed Fiber Optic Cable Plant (Insertion Loss, TIA OFSTP-14, OFSTP-7, ISO/IEC 61280, ISO/IEC 14763, etc.).
FOA Standard FOA-2: Testing Loss of Fiber Optic Cables, Single Ended (Insertion Loss, TIA FOTP-171, OFSTP-7, ISO/IEC 14763).
FOA Standard FOA-3: Measuring Optical Power (Transmitter and Receiver Power, FOTP-95, Numerous ISO/IEC standards).
FOA Standard FOA-4: OTDR Testing of Fiber Optic Cable Plant (TIA FOTP-8/59/60/61/78, ISO/IEC 14763, etc.).
FOA Standard FOA-7: Mode Conditioning for Testing Multimode Fiber Optic Cables (Mandrel wrap, encircled flux).

4.4.2   Fiber Optic Systems and Subsystems

The TIA website contains a comprehensive listing of Fiber Optic test Protocols also referred to as FOTPs. These are commonly known as "FOTPs" but are officially called EIA-455-x. A sample listing of these standards is given below and should not be deemed to be complete. Since the standards change continually, this list should be considered for reference purposes only. It would be prudent to visit the TIA website for a catalog of the latest versions.

FOTP-1 – Cable Flexing for Fiber Optic Interconnecting Devices (ANSI/TIA/EIA-455-1-B-98).
FOTP-2 – Impact Test Measurements for Fiber Optic Devices (ANSI/TIA/EIA-455-2-C-98).
FOTP-4 – Fiber Optic Component Temperature Life Test.
FOTP-5 – Humidity Test Procedure for Fiber Optic Components.

FOTP-6 – Cable Retention Test Procedure for Fiber Optic Interconnecting Devices.

FOTP-8 – Measurement of Splice or Connector Loss and Reflectance Using an OTDR.

FOTP-10 – Procedure for Measuring the Amount of Extractable Material in Coatings Applied to Optical Fibers (withdrawn April 1996).

FOTP-11 – Vibration Test Procedure for Fiber Optic Components and Cables.

FOTP-12 – Fluid Immersion Test for Fiber Optic Components.

FOTP-13 – Visual and Mechanical Inspection of Fibers, Cables, Connectors, and Other Devices.

FOTP-14 – Fiber Optic Shock Test (Specified Pulse).

FOTP-15 – Altitude/Immersion of Fiber Optic Components.

FOTP-16 – Salt Spray (Corrosion) Test for Fiber Optic Components.

FOTP-17 – Maintenance Aging of Fiber Optic Connectors and Terminated Cable Assemblies.

FOTP-18 – Acceleration Testing for Components and Assemblies.

FOTP-20 – IEC 60793-1-46 Optical Fibers – Part 1-46: Measurement Methods and Test Procedures – Monitoring of Changes in Optical Transmittance.

FOTP-21 – Mating Durability for Fiber Optic Interconnecting Devices.

FOTP-22 – Ambient Light Susceptibility of Components.

FOTP-23 – Air Leakage Testing for Fiber Optic Component Seals.

FOTP-24 – Water Peak Attenuation Measurement of Single-Mode Fibers.

FOTP-25 – Repeated Impact Testing of Fiber Optic Cables and Cable Assemblies.

FOTP-26 – Crush Resistance of Fiber Optic Interconnecting Devices.

FOTP-27 – Fiber Diameter Measurements.

FOTP-28 – Measuring Dynamic Strength and Fatigue Parameters of Optical Fibers by Tension.

FOTP-29 – Refractive Index Profile (Transverse Interference Method).

FOTP-30 – Frequency Domain Measurement of Multimode Optical Fiber Information Transmission Capacity (withdrawn May 2003).

FOTP-31 – Proof Testing Optical Fibers by Tension (2004) (R 2005).

FOTP-32 – Fiber Optic Circuit Discontinuities.

FOTP-33 – Optical Fiber Cable Tensile Loading and Bending Test.

FOTP-34 – Interconnection Device Insertion Loss Test.

FOTP-35 – Fiber Optic Component Dust (Fine Sand) Test.

FOTP-36 – Twist Test for Connecting Devices.

FOTP-37 – Low or High Temperature Bend Test for Fiber Optic Cable (ANSI/TIA/EIA-455-37-A-93) (R2000) (R 2005).

FOTP-39 – Fiber Optic Cable Water Wicking Test (ANSI/TIA/EIA-455-39B-99) (R 2005).

FOTP-40 – Fluid Immersion, Cables.

FOTP-41 – Compressive Loading Resistance of Fiber Optic Cables.

FOTP-42 – Optical Crosstalk in Components.

FOTP-43 – Output Near Field Radiation Pattern Measurement of Optical Waveguide Fibers.

FOTP-44 – Refractive Index Profile (Refracted Ray Method).

FOTP-45 – Microscopic Method for Measuring Fiber Geometry of Optical Waveguide Fibers.

FOTP-46 – Spectral Attenuation Measurement (Long Length Graded Index Optical Fibers).

FOTP-47 – Output Far Field Radiation Pattern Measurement.

FOTP-48 – Measurement of Optical Fiber Cladding Diameter Using Laser-Based Instruments (ANSI/TIA/EIA-455-48B-90) (R2000) (R 2005).

FOTP-49 – Measurement for Gamma Irradiation Effects on Optical Fiber and Cables.

FOTP-50 – Light Launch Conditions of Long-Length Graded-Index Optical Fiber Spectral Attenuation Measurements.

FOTP-51 – Pulse Distortion Measurement of Multimode Glass Optical Fiber Information Capacity.

FOTP-53 – Attenuation by Substitution Measurement for Multimode Graded-Index Optical Fibers of Fiber Assemblies Used in Long Length Communications Systems.

FOTP-54 – Mode Scrambler Requirements for Overfilled Launching Conditions to Multimode Fibers.

FOTP-55 – Methods for Measuring the Coating Geometry of Optical Fibers (withdrawn July 2000).

FOTP-56 – Test Method for Evaluating Fungus Resistance of Optical Waveguide Fiber.

FOTP-56 – Test Method for Evaluating Fungus Resistance of Optical Fiber and Cable (2004) (R 2005).

FOTP-57 – Preparation and Examination of Optical Fiber Endface for Testing Purposes.

FOTP-58 – Core Diameter Measurements (Graded Index Fibers).

FOTP-59 – Measurement of Fiber Point Defects Using an OTDR.

FOTP-60 – Measurement of Fiber or Cable Length Using an OTDR (superceded by TIA-455-133-A).

FOTP-61 – Method for Measuring the Effects of Nuclear Thermal Blast on Optical Waveguide Fiber.

FOTP-61 – Measurement of Fiber or Cable Attenuation Using an OTDR.

FOTP-62 – IEC 60793-1-43 Measurement Methods and Test Procedures – Numerical Aperture.

FOTP-62 – IEC 60793-1-47 Measurement Methods and Test Procedures – Macrobending Loss.

FOTP-63 – Torsion Test for Optical Fiber.

FOTP-64 – Procedures for Measuring Radiation-Induced Attenuation in Optical Fibers and Optical Cables.

FOTP-65 – Flexure Test for Optical Fiber.

FOTP-66 – Test Method for Measuring Relative Abrasion Resistance.

FOTP-67 – Procedure for Assessing High Temperature Exposure Effects on Optical Characteristics of Optical Fibers.

FOTP-67 – IEC 60793-1-51 Optical Fibers – Part 1-51: Measurement Methods and Test Procedures – Dry Heat.

FOTP-68 – Optical Fiber Microbend Test Procedure.

FOTP-69 – Test Procedure for Evaluating the Effect of Minimum and Maximum Exposure Temperature on the Optical Performance of Optical Fibers (ANSI/EIA/TIA-455-69A-91) (R2000).

FOTP-69 – Test Procedure for Evaluating the Effect of Minimum and Maximum Exposure Temperatures on the Optical Performance of Optical Fibers.

FOTP-70 – Procedure for Assessing High Temperature Exposure Effects on Mechanical Characteristics of Optical Fibers (superceded by ANSI/TIA-455-67-A).

FOTP-71 – Procedure to Measure Temperature-Shock Effects on Fiber Optic Components (ANSI/EIA/TIA-455-71-A-99).

FOTP-72 – Procedure for Assessing Temperature and Humidity Cycling Exposure Effects on Optical Characteristics of Optical Fibers.

FOTP-73 – Procedure for Assessing Temperature and Humidity Cycling Exposure Effects on Mechanical Characteristics of Optical Fibers.

FOTP-74 – IEC 60793-1-53 Optical Fibers – Part 1-53: Measurement Methods and Test Procedures – Water Immersion.

FOTP-75 – Fluid Immersion Aging Procedure for Optical Fiber Mechanical Properties.

FOTP-76 – Method of Measuring Dynamic Fatigue of Optical Fibers by Tension (withdrawn July 2000).

FOTP-77 – Procedures to Qualify a Higher-Order Mode Filter for Measurements on Single-Mode Fiber (withdrawn May 2003).

FOTP-78 – IEC 60793-1-40 Optical Fibers – Part 1-40: Measurement Methods and Test Procedures – Attenuation.

OTDR (ANSI/TIA/EIA-455-8-2000).

FOTP-80 – IEC 60793-1-44 Measurement Methods and Test Procedures – Cut-off Wavelength.

FOTP-81 – Compound Flow (Drip) Test for Filled Fiber Optic Cable (ANSI/EIA/TIA-455-81B-91) (R2000).

FOTP-82 – Fluid Penetration Test for Fluid-Blocked Fiber Optic Cable (1991).

FOTP-83 – Cable to Interconnecting Device Axial Compressive Loading.

FOTP-84 – Jacket Self-Adhesion (Blocking) Test for Cables.

FOTP-85 – Fiber Optic Cable Twist Test.

FOTP-86 – Fiber Optic Cable Jacket Shrinkage.

FOTP-87 – Fiber Optic Cable Knot Test.

FOTP-88 – Fiber Optic Cable Bend Test (ANSI/TIA/EIA-455-88-2001).

FOTP-89 – Fiber Optic Cable Jacket Elongation and Tensile Strength Test.

FOTP-91 – Fiber Optic Cable Twist-Bend Test.

FOTP-92 – Optical Fiber Cladding Diameter by Frizeau Interferometry (superceded by TIA-455-176-A).

FOTP-93 – Cladding Diameter by Non-Contacting Michelson Interferometry (withdrawn July 2000).

FOTP-94 – Fiber Optic Cable Stuffing Tubing Compression.

FOTP-95 – Absolute Optical Power Test for Optical Fibers and Cables (ANSI/TIA/EIA-455-95-A-2000) (R 2005).

FOTP-96 – Fiber Optic Cable Long-Term Storage Temperature Test for Extreme Environments.

FOTP-98 – Fiber Optic Cable External Freezing Test.

FOTP-99 – Gas Flame Test for Special Purpose Cable.

FOTP-100 – Gas Leakage Test for Gas Blocked Cable.

FOTP-101 – Accelerated Oxygen Test.

FOTP-102 – Water Pressure Cycling.

FOTP-104 – Fiber Optic Cable Cyclic Flexing Test.

FOTP-106 – Procedure for Measuring Near-Infrared Absorbance Coating Material (withdrawn September 2002).

FOTP-107 – Determination of Component Reflectance or Link/System Return Loss Using a Loss Test Set (2004).

FOTP-111 – IEC 60793-1-34 Optical Fibers – Part 1-34: Measurement Methods and Test Procedures – Fiber Curl.

FOTP-115 – Spectral Attenuation Measurement of Step-Index Multimode Optical Fibers.

FOTP-119 – Coating Geometry Measurement of Optical Fiber by Gray-Scale Analysis (withdrawn July, 2000).

FOTP-120 – Modeling Spectral Attenuation on Optical Fiber (superceded by TIA-455-78-B).

FOTP-122 – Polarization Mode Dispersion Measurement for Single Mode Optical Fibers by Stokes Parameter Evaluation.

FOTP-123 – Measurement of Optical Fiber Ribbon Dimensions (ANSI/TIA/EIA-455-123-2000) (R 2005).

FOTP-124 – Polarization-Mode Dispersion Measurement for Single-Mode Optical Fibers by Interferometry Method.

FOTP-124 – Polarization-Mode Dispersion Measurement for Single-Mode Optical Fibers by Interferometry.

FOTP-126 – Spectral Characterization of LEDs (ANSI/TIA/EIA-455-126-2000).

FOTP-127 – Spectral Characterization of Laser Diodes.

FOTP-130 – Elevated Temperature Life Test for Laser Diodes (ANSI/TIA/EIA-455-130-2001).

FOTP-131 – Measurement of Optical Fiber Ribbon Residual Twist (ANSI/TIA/ EIA-455-131-97).

FOTP-132 – Measurement of the Effective Area of Single-Mode Optical Fiber (ANSI/TIA/EIA-455-132-2001).

FOTP-133-A IEC 60793-1-22 Measurement Methods and Test Procedures – Length Measurement.

FOTP-141 – Twist Test for Optical Fiber Ribbons (ANSI/TIA/EIA-455-141-1999) (R 2005).

FOTP-157 – Measurement of Polarization Dependent (PDL) of Single-Mode Fiber Optic Components (ANSI/TIA/EIA-455-157-1995) (R2000).

FOTP-158 – Measurement of Breakaway Frictional Face in Fiber Optic Connector Alignment Sleeves.

FOTP-160 – IEC 60793-1-50 Optical Fibers – Part 1–50: Measurement Methods and Test Procedures – Damp Heat (Steady State).

FOTP-161 – Procedure for Assessing Temperature and Humidity Exposure Effects on Mechanical Characteristics of Optical Fibers (superceded by ANSI/TIA-455-160-A).

FOTP-162 – Optical Fiber Cable Temperature-Humidity Cycling.

FOTP-164 – Measurement of Mode Field Diameter by Far-Field Scanning (Single-Mode).

FOTP-165 – Single-Mode Fiber Diameter by Near-Field Scanning Technique (withdrawn July 2000).

FOTP-166 – Transverse Offset Method.

FOTP-167 – Mode Field Diameter Measurement – Variable Aperture Method in Far-Field (withdrawn September 2002).

FOTP-168 – Chromatic Dispersion Measurement of Multimode Graded-Index and Single-Mode Optical Fibers by Spectral Group Delay Measurement in the Time Domain (superceded by TIA-175-B).

FOTP-169 – Chromatic Dispersion Measurement of Single-Mode Optical Fibers by the Phase-shift Method (superceded by TIA-455-175-B).

FOTP-170 – Cable Cutoff Wavelength of Single-Mode Fiber by Transmitted Power.

FOTP-171 – Attenuation by Substitution Measurement for Short-Length Multimode Graded-Index and Single-Mode Optical Fiber Cable Assemblies (ANSI/TIA/EIA-455-171-A-2001).

FOTP-172 – Flame Resistance of Firewall Connector.

FOTP-173 – Coating Geometry Measurement of Optical Fiber, Side-View Method.

FOTP-174 – Mode Field Diameter of Single-Mode Fiber by Knife-Edge Scanning in Far-Field.

FOTP-175 – IEC 60793-1-42 Measurement Methods and Test Procedures – Chromatic Dispersion.

FOTP-176 – A IEC 60793-1-20 Measurement Methods and Test Procedures – Fiber Geometry.

FOTP-177 – Numerical Aperture Measurement of Graded-Index Fiber.

FOTP-178 – IEC 60793-1-32 Optical Fibers – Part 1–32: Measurement Methods and Test Procedures – Coating Strippability.

FOTP-179 – Inspection of Cleaved Fiber End Faces by Interferometry.

FOTP-180 – Measurement of the Optical Transfer Coefficients of a Passive Branching Device (Coupler).

FOTP-181 – Lightning Damage Susceptibility Test for Fiber Optic Cables with Metallic Components (ANSI/TIA/EIA-455-181-92) (R2001).

FOTP-183 – Hydrogen Effects on Optical Fiber Cable (ANSI/TIA/EIA-455-183-2000) (R 2005).

FOTP-184 – Coupling Proof Overload Test for Fiber Optic Interconnecting Devices (ANSI/TIA/EIA-455-184-91) (R95) (R99).

FOTP-185 – Strength of Coupling Mechanism for Fiber Optic Interconnecting Devices (ANSI/TIA/EIA-455-185-91) (R95) (R99).

FOTP-186 – Gauge Retention Force Measurement for Fiber Optic Components (2004).

FOTP-187 – Engagement and Separation Force Measurement of Fiber Optic Connector Sets (2004).

FOTP-188 – Low-Temperature Testing for Components.

FOTP-189 – Ozone Exposure Test for Fiber Optic Components.

FOTP-190 – Low Air Pressure (High Altitude) Test for Components.

FOTP-191 – IEC 60793-1-45 Optical Fibers – Part 1–45: Measurement Methods and Test Procedures – Mode Field Diameter.

FOTP-193 – Polarization Crosstalk Method for Polarization Maintaining Optical Fiber and Components.

FOTP-194 – Measurement of Fiber Pushback in Optical Connectors (ANSI/TIA/EIA-455-194-99).

FOTP-195 – IEC 60793-1-21 Optical Fibers – Part 1–21: Measurement Methods and Test Procedures – Coating Geometry.

FOTP-196 – Guideline for Polarization-Mode Measurement in Single-Mode Fiber Optic Components and Devices (ANSI/TIAEIA-455-196-99).

FOTP-197 – Differential Group Delay Measurement of Single-Mode Components and Devices by the Differential Phase Shift Method (ANSI/TIA/EIA-455-197-2000).

FOTP-198 – Measurement of Polarization Sensitivity of Single-Mode Fiber Optic Components by Matrix Calculation Method.

FOTP-199 – In-line Polarization Crosstalk Measurement Method for Polarization – Maintaining Optical Fibers Components and Systems.

FOTP-200 – Insertion Loss of Connectorized Polarization-Maintaining Fiber or Polarizing Fiber Pigtailed Devices and Cable Assemblies (ANSI/TIA/EIA-455-200-2001).

FOTP-201 – Return Loss of Commercial Polarization – Maintaining Fiber or Polarizing Fiber Pigtailed Devices and Cable Assemblies (ANSI/TIA/EIA-455-201-2001).

FOTP-203 – Launched Power Distribution Measurement Procedure for Graded-Index Multimode Fiber Transmitters (ANSI/TIA/EIA-455-203-2001).

FOTP-204 – Measurement of Bandwidth on Multimode Fiber (ANSI/TIA/EIA-455-204-2000).

FOTP-206 – IEC 61290-1-1 Optical Fiber Amplifiers – Basic Specification Part 1-1: Test Methods for Gain Parameters – Optical Spectrum Analyzer (ANSI/TIA/EIA-455-206-2000).

FOTP-207 – IEC 61290-1-2 Optical Fiber Amplifiers – Basic Specification Part 102: Test Methods for Gain Parameters – Electrical Spectrum Analyzer (ANSI/TIA/EIA-455-207-2000).

FOTP-208 – IEC 61290-1-3 Optical Fiber Amplifiers – Basic Specification Part 1-3: Test Methods for Gain Parameters – Optical Power Meter (ANSI/TIA/EIA-455-208-2000).

FOTP-209 – IEC 61290-2-1 Optical Fiber Amplifiers – Basic Specification Part 2-1: Test Methods for Optical Power Parameters – Optical Spectrum Analyzer (ANSI/TIA/EIA-455-209-2000).

FOTP-210 – IEC 61290-2-2 Optical Fiber Amplifiers – Basic Specification Part 2-2: Test Methods for Optical Power Parameters – Electrical Spectrum Analyzer (ANSI/TIA/EIA-455-210-2000).

FOTP-211 – IEC 61290-2-3 Optical Fiber Amplifiers – Basic Specification Part 2-3: Test Methods for Optical Power Parameters – Optical Power Meter (ANSI/TIA/EIA-455-211-2000).

FOTP-212 – IEC 61290-6-1 Optical Fiber Amplifiers – Basic Specification Part 6-1: Test Methods for Pump Leakage Parameters – Optical Demultiplexer (ANSI/TIA/EIA-455-212-2000).

FOTP-213 – IEC 61290-7-1: Optical Fiber Amplifiers – Basic Specification Part 7-1: Test Methods for Out-of-Band Insertion Losses – Filtered Optical Power Meter (ANSI/TIA/EIA-455-213-2000).

FOTP-214 – IEC 61290-1 Optical Fiber Amplifiers – Part 1: Generic Specification (ANSI TIA/EIA-455-214-2000).

FOTP-218 – Measurement of Endface Geometry of Optical Connectors.

FOTP-219 – Multifiber Ferrule Endface Geometry Measurement.

FOTP-220 – Differential Mode Delay Measurement of Multimode Fiber in the Time Domain (superseded by TIA-455-220-A).

FOTP-220 – Measurement of Minimum Modal Bandwidth of Multimode Fiber Using Differential Mode Delay.

FOTP-221 – IEC 61290-5-1 Optical Fiber Amplifiers – Basic Specification – Part 5-1: Test Method for Reflectance Parameters – Optical Spectrum Analyzer.

FOTP-222 – IEC 61290-3 – Optical Fiber Amplifiers – Basic Specification – Part 3: Test Methods for Noise Parameters.

FOTP-223 – IEC 61291-2 – Optical Fiber Amplifiers – Part 2: Digital Applications – Performance Specification Template.

FOTP-224 – IEC 61744 Calibration of Fiber Optic Chromatic Dispersion Test Sets.

FOTP-225 – IEC 61745 End-Face Image Analysis Procedure for the Calibration of Optical Fiber Geometry Test Sets.

FOTP-225 – IEC 61745, Ed. 1.0 (1998-08): End Face Image Analysis Procedure for the Calibration of Optical Fiber Geometry Test Sets.

FOTP-226 – IEC 61746 Calibration of Optical Time-Domain Reflectometers (OTDR's).

FOTP-227 – IEC 61300-3-24 Fiber Optic Interconnecting Devices and Passive Components – Basic Test and Measurement Procedures – Part 3–24: Examination and Measurements – Keying Accuracy of Optical Connectors for Polarization Maintaining Fiber.

FOTP-228 – Relative Group Delay and Chromatic Dispersion Measurement of Single-Mode Components and Devices by the Phase Shift Method.

FOTP-229 – Optical Power Characterization.

FOTP-231 – IEC 61315 Calibration of Fiber-Optic Power Meters.

FOTP-234 – IEC 60793-1-52 Optical Fibers – Part 1–52: Measurement Methods and Test Procedures – Change of Temperature.

FOTP-239 – Fiber Optic Splice Loss Measurement Methods.

The following is a list of OFSTPs.

OFSTP-2 – Effective Transmitter Output Power Coupled into Single-Mode Fiber Optic Cable.

OFSTP-3 – Fiber Optic Terminal Equipment Receiver Sensitivity and Maximum Receiver Input.

OFSTP-4 – Optical Eye Pattern Measurement Procedure.

OFSTP-7 – Measurement of Optical Power Loss of Installed Single-Mode Fiber Cable Plant (2003).

OFSTP-11 – Measurement of Single-Reflection Power Penalty for Fiber Optic Terminal Equipment OFSTP-15 Jitter Tolerance Measurement.

OFSTP-14 – IEC-61280-4-1 (2011) (This ISO/IEC document is written for fast MM networks and may not be useful for other MM networks, so the prior version of OFSTP-14 is still considered valid), Optical Power Loss Measurement of Installed Multimode Fiber Cable Plant (1998) (R2003).

OFSTP-16 – Jitter Transfer Function Measurement.

OFSTP-17 – Output Jitter Measurement.

OFSTP-18 – Systematic Jitter Generation Measurement.

OFSTP-19 – Optical Signal-to-Noise Ratio Measurement Procedures for Dense Wavelength-Division Multiplexed Systems (ANSI/TIA/EIA-526-19-2000).

OFSTP-27 – Procedure for System Level Temperature Cycle Endurance Test.

OFSTP-28 – IEC-61290-1-2: Basic Spec For Optical Fiber Amplifiers Test Methods Part 1: Test Methods For Gain Parameters – Sect. 2: Electrical Spectrum Analyzer Test Method.

OFSTP-29 – IEC-61290-1-3: Basic Spec For Optical Fiber Amplifiers Test Methods Part 1: Test Methods For Gain Parameters – Sect. 3: Optical Power Meter Test Method.

OFSTP-30 – IEC-61290-2-1: Basic Specification for Optical Fiber Amplifiers Test Methods – Part 2: Test Methods for Spectral Power Parameters – Section 2 – Optical Spectrum Analyzer Test Method.

In addition, the following is a list of Electronics Industry Association Standards (EIA).

EIA-458 – B Standard Optical Fiber Material Classes and Preferred Sizes.

EIA-472 – General Specification for Fiber Optic Cable.

EIA-472A – Sectional Specification for Fiber Optic Communication Cables for Outside Aerial Use.

EIA-472B – Sectional Specification for Fiber Optic Communication Cables for Underground and Buried Use.

EIA-472C – Sectional Specification for Fiber Optic Communication Cables for Indoor Use.

EIA-472D – Sectional Specification for Fiber Optic Communication Cables for Outside Telephone Plant Use.

EIA-4750000 – B Generic Specification for Fiber Optic Connectors.

EIA-475COOO – Sectional Specification for Type FSMA Connectors.

EIA-475CAOO – Blank Detail Specification for Optical Fiber and Cable Type FSMA, Environmental Category I EIA-475CBOO – Blank Detail Specification Connector Set for Optical Fiber and Cables Type FSMA, Environmental Category 11.

EIA-475CCOO – Blank Detail Specification Connector Set for Optical Fiber and Cables Type FSMA, Environmental Category III.

EIA-475EOOO – Sectional Specification for Fiber Optic Connectors Type BFOC/2.5.

EIA-475EAOO – Blank Detail Specification for Connector Set for Optical Fiber and Cables, Type BFOC/2.5, Environmental Category I.

EIA-475EBOO – Blank Detail Specification for Connector Set for Optical Fiber and Cables, Type BFOC/2.5, Environmental Category 11.

EIA-475ECOO – Blank Detail Specification for Connector Set for Optical Fiber and Cables, Type BFOC/2.5, Environmental Category III.

TIA TIA-4920000 – B EN-Generic Specification for Optical Fibers.

TIA TIA-492A000 – A EN-Sectional Specification for Class Ia Graded-Index Multimode Optical Fibers.

TIA TIA-492AA00 – A EN-Blank Detail Specification for Class Ia Graded-Index Multimode Optical Fibers.

TIA TIA-492AAAA – B EN-Detail Specification for 62.5-μm Core Diameter/125-μm Cladding Diameter Class Ia Graded-Index Multimode Optical Fibers.

TIA TIA-492AAAB – A EN-Detail Specification for 50-μm Core Diameter/125-μm Cladding Diameter Class Ia Graded-Index Multimode Optical Fibers.

TIA TIA-492AAAC – B EN-Detail Specification for 850-nm Laser-Optimized 50-μm Core Diameter/125-μm Cladding Diameter Class Ia Graded-Index Multimode Optical Fibers.

TIA TIA-492AAAD – EN-Detail Specification for 850-nm Laser-Optimized 50-μm Core Diameter/125-μm Cladding Diameter Class la Graded-Index Multimode Optical Fibers Suitable for Manufacturing OM4 Cabled Optical Fiber.

TIA TIA-492C000 – EN-Sectional Specification for Class IVa Dispersion-Unshifted Single-Mode Optical Fibers.

TIA TIA-492CA00 – EN-Blank Detail Specification for Class IVa Dispersion-Unshifted Single Mode Optical Fibers.

TIA TIA-492CAAA – EN-Detail Specification for Class IVa Dispersion-Unshifted Single-Mode Optical Fibers.

TIA TIA-492CAAB – EN-Detail Specification for Class IVa Dispersion-Unshifted Single-Mode Optical Fibers with Low Water Peak.

TIA TIA-492E000 – EN-Sectional Specification for Class IVd Nonzero-Dispersion Single-Mode Optical Fibers for the 1550 nm Window.

TIA TIA-492EA00 – EN-Blank Detail Specification for Class IVd Nonzero-Dispersion Single-Mode Optical Fiber for the 1550 nm Window.

EIA-5390000 – Generic Specification for Field Portable Polishing Device for Preparation Optical Fiber.

EIA-5460000 – Generic Specification for a Field Portable Optical Inspection Device, Combined EIA-NECQ Specification.

EIA-546A000 – Sectional Specification for a Field Portable Optical Microscope for Inspection of Optical Waveguide and Related Devices.

EIA-587 – Fiber Optic Graphic Symbols.

EIA-590 – Standard for Physical Location and Protection of Below-Ground Fiber Optic Cable Plant.

EIA-598 – Color Coding of Fiber Optic Cables.

IEC Standards

IEC/ISO Standards can be searched at http://webstore.ansi.org/RecordDetail.aspx?sku=IEC+Catalog Standard Number.

IEC 60793 – Optical fibers.

IEC 60794 – Optical fiber cables.

IEC 60869 – Fiber optic attenuators.
IEC 60874 – Connectors.
IEC 60875 – Fiber optic branching devices.
IEC 60876 – Fiber optic spatial switches.
IEC 61073 – Splices for optical fibers and cables.
IEC 61202 – Fiber optic isolators.
IEC 61274 – Fiber optic adaptors.
IEC 61280 – Fiber optic communication subsystem basic test procedures.
IEC 61281 – Fiber optic communication subsystems.
IEC 61282 – Fiber optic communication system design guides.
IEC 61290 – Optical amplifier test methods.
IEC 61291 – Optical amplifiers.
IEC 61292 – TRs Optical amplifiers technical reports.
IEC 61300 – Test and measurement.
IEC 61313 – Fiber optic passive components.
IEC 61314 – Fiber optic fan-outs.
IEC 61751 – Laser modules used for telecommunication.
IEC 61753 – Fiber optic interconnecting devices and passive components performance standard.
IEC 61754 – Fiber optic connector interfaces.
IEC 61755 – Fiber optic connector optical interface.
IEC 61756 – Fiber management system.
IEC 61757 – Fiber optic sensors.
IEC 61977 – Fiber optic filters.
IEC 61978 – Fiber optic passive dispersion compensators.
IEC 62005 – Reliability.
IEC 62007 – Semiconductor optoelectronic devices.
IEC 62074 – Fiber optic WDM devices.
IEC 62077 – Fiber optic circulators.
IEC 62099 – Fiber optic wavelength switches.
IEC 62134 – Fiber optic enclosures.
IEC 62148 – Fiber optic active components and devices – package and interface standards.
IEC 62149 – Fiber optic active components and devices – performance standards.
IEC 62150 – Fiber optic active components and devices – test and measurement procedures.
IEC 62343 – Dynamic modules.
IEC 61757-3-1 is a standard that defines these parameters and can be a good informative document for further reading.

The IEC Standard 61757-2-2 defines detailed specifications for distributed temperature measurement by a fiber optic sensor. DTS includes the use of Raman, Brillouin, and Rayleigh scattering effects. Raman and Rayleigh scattering-based measurements are performed with a single-ended fiber configuration only. Brillouin scattering-based measurements are performed with a single-ended fiber or fiber loop configuration. The technique accessible from both sides at the same time (e.g. Brillouin optical time domain analysis, BOTDA) is referred to here as a loop configuration. Generic specifications for fiber optic sensors are defined in the IEC 61757 (2016). This part of IEC 61757 specifies the most important DTS performance parameters and defines the procedures for their determination. In addition to the group of performance parameters, a list of additional parameters has been defined to support the definition of the measurement specifications and their associated test procedures. The definitions of these additional parameters are provided for informational purposes and should be included with the sets of performance parameters. A general test setup is defined in which all parameters can be gathered through a set of tests. The specific tests are described within the clause for each measurement parameter. This general test setup is depicted and described in Clause 4 of the standard along with a list of general information that should be documented based upon the specific DTS instrument and test setup used to measure these parameters as per IEC 61757-2-2. Annex A provides a blank performance parameter table which should be used to record the performance parameter values for a given DTS instrument and chosen optical test setup configuration. Annex B provides guidelines for optional determination of point defect effects.

4.5 Optical Connectors

High-quality optical fiber connections and fusion splices are important because they affect light attenuation and thereby the ability of DTS systems to detect faint backscattered light, which in turn affects measurement accuracy, spatial resolution, calculation time, and so on. Ideally, all connections should be made with high-quality fusion splices; however, connections to the DTS equipment and calibrators must be removable. Following is a summary of the main issues related to the optical connections to patch panels and DTS equipment.

Typically, optical fibers from the field are terminated in a patch panel or splice enclosure close to the DTS equipment, using fusion splices or mechanical connectors. Alternatively, pigtail leads with a mechanical connector on the DTS end can be fusion spliced directly onto the field cables. In either case, it is important that all connectors are compatible with the DTS equipment, as well as those used on any patch panel or splice enclosure. The optical fiber type used in the pigtails must also be the same as used for the field fiber (MM or SM) and specified by the DTS equipment provider.

Pigtails and their connectors must meet the requirements of the IEC 61300 "Fiber optic interconnecting devices and passive components" and IEC 60874-14 "Connectors for optical fibers and cables."

There are many types of optical connectors used in the industry today which are listed in Figure 4.22.

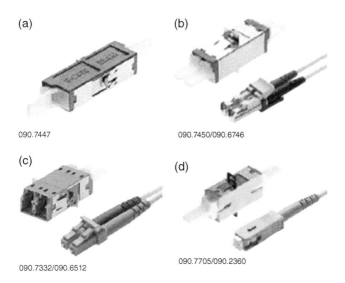

(a)

090.7447

(b)

090.7450/090.6746

(c)

090.7332/090.6512

(d)

090.7705/090.2360

Figure 4.22 (a) Simplex E-2000; (b) duplex small form factor; (c) LC simplex and LC duplex; (d) SC-simplex and duplex form factor. (*Source:* courtesy R&M Systems.)

Figure 4.23 Type of ends on a ferrule connector.

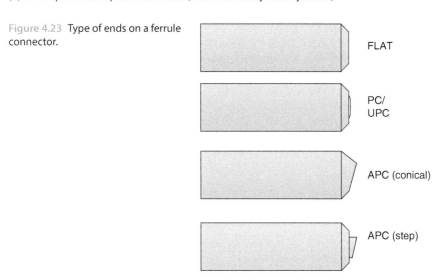

FLAT

PC/ UPC

APC (conical)

APC (step)

Figure 4.23 shows a cross-section of the end section of a typical ferrule on some connectors.

The flat end is where an optical fiber is cleaved with the cleave occurring at 90° to the fiber axis as illustrated. The designation PC/UPC is referred to when the same cleaved end is either polished (PC) to ultra-polished (UPC). The types of connector ends will present little reflection when connectors are made. They are susceptible to even the minor scratches; hence, they require great care in handling. The angled physical contact (APC) type end refers to an angle-polished connector that is conical at a cleave angle of about 8°. Experience has shown that these are the types of connector ends for the DTS temperature measurements.

The connector geometry is the profile of the end face of a ferrule once polished. Ideally, the profile of a connector once polished should conform to an established standard thereby ensuring intermateability between different connector manufacturers. Unfortunately, some manufacturers and installers do not care about the end face geometry.

In practice, polishing single-mode connectors manually or without regular interferometric inspection can be catastrophic for modern high-speed data systems. The following section outlines some typical requirements for such intermateability and alerts the reader their importance.

- Apex Offset is defined as the distance (μm) between the core's central axis of the fiber and the highest point on the end face of the ferrule. In an ideal connector there should be no offset but it is accepted that one should try and achieve an offset that is smaller than 50 μm.
- Radius of curvature is the distance from the cable axis to the dome of the ferrule and should be 10–25 mm for a polished connector and 5–12 mm for an angle-polished connector. Ideal connectors would have a radius of curvature ranging from 17.5 to 8.5 mm.
- Fiber height is the position of the fiber relative to the fiber end face (expressed in nm). The usual way to measure fiber height is by spherical simulation (see the last diagram in Figure 4.24). Negative values refer to undercut (fiber recessing the ferrule) and positive values refer to protrusion (fiber exceeding the ferrule). Fiber height should be smaller than ±50 nm. The ideal connector would have the fiber at 0 nm. An alternative method to calculate protrusion or undercut is the planar method. This represents the difference in height between the fiber center and the average height of the ferrule adjacent to the fiber perimeter.

The visual inspection is a critical step in the manufacturing process of a connector, a ferrule, or any other component destined to transfer light from one fiber to another.

In the manufacturing cycle, this step is necessary before the interferometric and qualification tests upon delivery. In use, a visual inspection is also very desirable before connection. The visual inspection can be done either by microscopic inspection which consists of enlarging the end of the ferrule and subjectively interpreting its quality, cleanliness, and so on, according to established criteria. Alternatively, a computerized inspection that is automated according to the established criteria can also be adopted. The best way to get repetitive results during inspections is to define the most precise possible criteria. The prerequisite elements to the success of an inspection lie in choosing an approved connector, ensuring good polishing quality, and

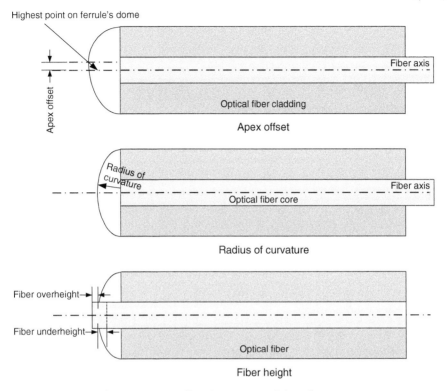

Figure 4.24 Type of issues that can affect the intermateability of connectors.

good cleaning. A microscopic inspection is usually done with a magnification of at least 100×.

The following section describes some of the requirements for optical performance requirements when connectors are employed. Using connectors helps intermateability and also offers flexibility to the user to interchange connections, move from different channels, testing, calibrations, and so on. Notwithstanding these advantages, the use of connectors can introduce undesirable optical losses and reduce the optical budgets required for accurate temperature measurements. The following section provides a short discussion on some of the parameters that introduce such losses.

Insertion loss is the loss resulting from the introduction of a passive component (connector, fiber, etc.) in a system. For an optic cable, this loss is specified at a certain wavelength (nm) for a single connector (such as E-2000APC, LC). For short lengths, the loss of the fiber is ignored (single mode: 0.5 dB/km; multimode: 3.5 dB/km). The loss of the second connector is also ignored because there is no PC, but aligned instead with a detector on a test bench. The insertion loss of a connector is due mainly to the manner in which the two fibers are aligned. Any variation in the fiber angle or axis will lead to an insertion loss. Thus, it is important to select the adequate connector in order to align both fibers as precisely as possible. A 125-μm connector will align a 125-μm fiber more precisely than a 126-μm or 127-μm connector. The manufacturing and polishing quality is just as important to ensure minimum insertion loss. It is common in a well-defined test bed to achieve insertion losses as low as 0.01–0.03 dB.

The other parameter that can affect accuracy is back reflection. This occurs when the incident light wave is reflected when two fibers are mated and which returns in the opposite direction; it is expressed in dBm of the reflected wave. Back reflection is expressed in negative terms, whereas return loss is expressed in positive terms. Back reflection is directly associated with the polishing quality of the connector as well as the geometry of the ferrule. The introduction of a PC has improved the performance of connectors significantly. Polishing procedures have also been improved (SPC and UPC). More recently, the introduction of an 8° APC ferrule has decreased reflection by at least 65 dB. It is important to note that the APC connector is not compatible with other types of connectors because of its angle. In the case of single-mode connectors, back reflection must be minimized especially to protect sensitive active components such as the laser sources. For multimode connectors, little attention was usually paid to the reflection, which often varies between 15 and 30 dB. Today, some manufacturers use the same procedures for all their connectors, whether that are single or multimode. Therefore, performance is optimized for all applications. Table 4.5 summarizes the possible levels of back reflection for various connector types.

There are many different connector types; however, the one most commonly used for the DTS applications is the E2000/APC with an 8° angle. An alternative is E2000/UPC (ultraphysical contact). Some DTS suppliers use either E2000/APC or FC/PC connectors, depending on their DTS system model number. Others use SC/APC connectors. Traditionally, for long-distance communication systems with optical fibers, many utilities and their telecom installers use ST connectors (straight tip or bayonet type) and most of the

Table 4.5 Comparative levels of back reflection that may be encountered based on connector type.

Type of polished connector	Description	Back reflection
PC	Physical contact	−30 dB
SPC	Superphysical contact	−45 dB
UPC	Ultraphysical contact	−55 dB
APC	Angled polish contact	−65 dB

(a)   (b)

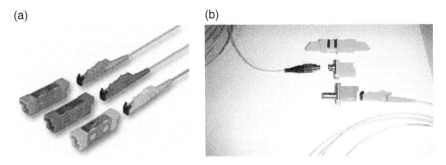

Figure 4.25 (a) Type E2000 connector system; (b) hybrid mating adapters between SC/APC and E2000/Angle Polished Connector (APC).

testing equipment (such as OTDR) accommodates only such connectors. This hinders the performance testing of the installed cables for DTS applications as these fibers are terminated with E2000/APC or SC/APC style connectors. In the past, the test personnel used mechanical splices to connect to the sensing fiber to ascertain the quality of the fiber and confirm that during its installation the optical integrity was not compromised. Fortunately, to address these incompatibilities, hybrid mating adapters have become available. Figure 4.25b shows photographs of such hybrid mating adapters allowing interconnection between SC/APC and E2000/APC without significantly increasing optical attenuation. Photos of E2000 connectors on pigtail leads and hybrid mating adapters are shown in Figure 4.25a and b.

Additional information is available from key suppliers, who developed and trademarked the de facto E2000 connector standard.

Unfortunately, many DTS system suppliers do not describe the type of connectors required for their systems in their data sheets, so care must be taken to ensure coordination and standardization when selecting new DTS system providers and coordinating with field installers of the optical fiber systems. There are also differing philosophies between DTS system manufacturers about the performance of the E2000/APC connectors versus the SC/APC models. Connector standardization will continue to be a challenge; however, adapters are available.

It is noteworthy that conventional OTDR measurements at 850 nm are good in isolating gross splicing and fiber defects, but they are inadequate to confirm FO quality at DTS wavelengths. In order to accurately and meaningfully measure attenuation, the wavelengths used should be closer to the DTS laser wavelength (which varies with the DTS system used), but generally around 1310 nm. Typically, for temperature applications, one can use SM and/or MM fibers. Industry standards are developed and all DTS fibers should meet minimum international telecommunication user standards (ITUs).

As mentioned earlier, for short-distance applications (<30 km), MM fibers are often applied. For long-distance applications (>30 km), SM fibers are usually used. Single mode can be used for short- or long-distance measurements while the use of MM fiber for temperature measurement for long-distance application affects measurement's capability and accuracy. Note that this threshold depends a lot on optical budget and the technology being used meaning that SM or MM fibers could be used, respectively, on shorter or longer monitoring ranges. In the MM fiber construction, it is quite common that most DTS suppliers will prefer using 50/125 μm instead of 62.5/125 μm. The choice on the numbers of fibers to use for temperature measurements may vary depending on the reliability of service required and the type of measurement implemented (single vs. double ended/loop).

4.6 Utility Practice for Testing of Optical Fibers

Before an optical fiber is installed, the following tests are typically performed in four stages:

1) FO cable manufacturer testing: the fiber optic testing generally follows ITU standards: mechanical (cold bend test), optical (integrity of the fiber, attenuation test OTDR, bit error rate test GHz-km), electrical (dielectric test), and geometrical verification.
2) FO cable tests following transportation: these tests are typically undertaken to confirm that delivered construction of ADSS or other type of FO cable is intact and has not suffered any mechanical or optical damage during transportation to the end-user job site.
3) FO cable tests following installation: these tests are undertaken to confirm that installed optical cable system's integrity has not been compromised. Examples: fiber break due to excessive pulling force exerted on the cable or negotiating bends along the corridor and violating the manufacturers' minimum bending radius criteria. Such tests also need to be conducted in order to confirm that the splicing/terminating operations have been done with adequate care to ensure losses are at the minimum acceptable for temperature monitoring.
4) FO cable system commissioning test: these are end-to-end tests to confirm that the system is fully operational, accurate, and reliable before placing into services.

It is of utmost importance to confirm these tests are done with considerable care and process to ensure that installation is in compliance with all the site acceptance test requirements. Any corrections required following these steps makes it exceedingly difficult because of special outage requirements to be obtained for undertaking this work and the associated extra costs that may be incurred.

In the case of embedded fiber, the power cable manufacturer goes through extra tests to confirm that the delivered optical fiber embedded in the cable has not suffered further damage during power cable production and installation. The tests are very similar to the one listed above.

The tests are described in more detail in other parts of this book.

4.7 Aging and Maintenance

There is no documented evidence of the long-term deterioration of performance in optical fibers at normal temperatures experienced in a cable system application (<90 °C). The telecom industry recognized significant attenuation between 850 and 1310 nm (attributed to water peak issue). Since the optical glass fiber is surrounded by many layers of polymeric materials, gels, and so on, the maximum temperature of the operation for this fiber will be dictated by the material properties.

The longevity of optical fibers embedded within the power cables remains unknown. Thermomechanical bending, the constant pulsing of the laser, temperature, and moisture can affect the remaining lifetime and attenuation of the OF.

The following are some known aging mechanisms for optical fibers:

- Hydrogen darkening effect is caused by hydrogen diffusing through the steel tube wall and affecting the attenuation in the fiber. This phenomenon is not known to occur in power cable temperature monitoring applications because temperatures rarely remain sustained at 250 °C and above. This temperature level is reached in a power cable during fault conditions which are of very short durations. Should hydrogen darkening happen, the maximum induced loss due to hydrogen should be less than 0.003 dB/km at 1550 nm over the 25-year design life of the system. Such hydrogen darkening-induced attenuation may affect the temperature accuracy of the DTS systems. If the end user does require resistance to hydrogen darkening, they will need to purchase a different and more expensive type of FIMT (example: carbon coated fibers or hydrogen scavengers in the protective gel inside the FIMT around the fibers).
- Radiation effects on the FIMT performance in power cable temperature monitoring applications are rare unless these power cables are placed in a radiation environment. Should the end user require resistance to radiation effects, the maximum radiation-induced losses for the fiber optic cable should not exceed 0.001 dB/km after 25 years. If the end user does require resistance to radiation effects, they will need to purchase a different and more expensive type of fiber optic cable.

- Optical fibers are usually maintenance free and, if left undisturbed, have been shown to perform reliably. Problem of inaccurate measurements typically are related to electronics, laser power losses, or the fiber being physically damaged. Quality of fiber attenuation and optical budget are all examined prior to commissioning the system to confirm any potential anomalies in the optical path. Anomalies are usually the result of mechanical deformation, improper handling, damage during cable operation (thermomechanical forces), or poor connectors.

5

Types of Power Cables and Cable with Integrated Fibers

This chapter addresses practical issues related to integration of the fiber optic element into power cables.

5.1 Methods of Incorporating DTS Sensing Optical Fibers (Cables) into Power Transmission Cable Corridors

5.1.1 Integration of Optical Cable into Land Power Cables

When utilities install brand new circuits, they have the choice of specifying a power cable system with an embedded optical fiber that can be placed in a (stainless) steel, copper, or plastic tube below the outer sheath. The sheath designs will govern the choice of the encasing tube. Figure 5.1 shows an example of an optical fiber encased in a stainless steel tube placed at the opposite ends of the cable circumference. For mechanical protection, the stainless steel tube is sandwiched between two copper wires and taped to the semiconducting cushion layers. This degree of protection is needed as the sheath is an extruded corrugated aluminum construction and sees considerable mechanical forces during manufacture. Typically, depending on the internal diameter of the stainless steel tube (usually, 1.6 mm outer diameter), one to four fibers are incorporated into the tube. The tube is also filled with some form of water swellable powder or thixotropic gel to inhibit longitudinal moisture migration. This encasing tube can either be stainless steel or copper. A stainless steel tube offers better mechanical protection but can remain stiff and difficult to cut. A copper tube, on the other hand, offers some flexibility but is comparatively weaker than stainless steel resulting in poor crush strength; a consideration important when the power cables are being placed in ducts and have to make a 90° bend around a street corner. Even though for the purposes of temperature measurement a single fiber is adequate, because of the fragility of the

Distributed Fiber Sensing and Dynamic Rating of Power Cables, First Edition.
Sudhakar Cherukupalli and George J. Anders.
© 2020 by The Institute of Electrical and Electronics Engineers, Inc.
Published 2020 by John Wiley & Sons, Inc.

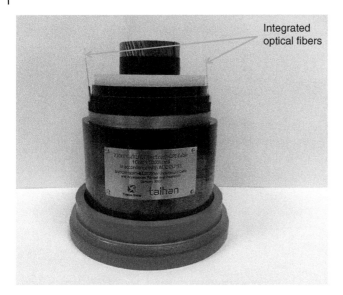

Figure 5.1 Photograph of a power cable system with embedded fibers located below a lead sheath. (*Source:* photo courtesy of Taihan Corporation, Korea.)

composite assembly, quite often redundancy is provided by introducing three or more optical fibers, ideally distributed around the cable circumference as shown in the examples in Figures 5.1 and 5.2.

5.1.2   Integration of Optical Cable into Submarine Power Cables

Figures 5.3 and 5.4 show how the optical fibers are placed in a stainless steel tube and supported by polyethylene terephthalate rounds as mechanical protection. The same have also been used as fillers to maintain the roundness of the cable construction and mechanically support the galvanized steel armor.

In three-core submarine cables, fiber optic sensors are sometimes installed in the interstices between the cores as shown in Figure 5.5.

5.1.3   Other Types of Constructions

Sometimes, because of concern about fiber damage during the manufacture of the cable, stainless steel encased fiber cables may also be attached mechanically with tie wraps on the surface of the cable as shown in Figure 5.6. This is an example of a submarine cable where the optical fiber is attached from the cable's termination to the low–low water mark.

Quite often when bundling optical fibers under an extruded sheath, the complexity of production increases and to avoid damage to the stainless steel cable, it is helically wrapped around the cable core.

In the case the end user is interested in monitoring an existing corridor and isolating any hot spots along the route, they have the option of placing an

Figure 5.2 Photograph of the cross-section of a 230 kV XLPE insulated power cable with a laminated sheath and the optical fiber installed in stainless steel tubes located at four positions around the circumference. (*Source:* photo courtesy of Taihan Corporation, Korea.)

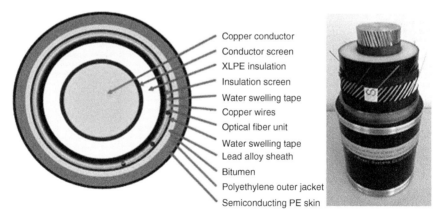

Copper conductor
Conductor screen
XLPE insulation
Insulation screen
Water swelling tape
Copper wires
Optical fiber unit
Water swelling tape
Lead alloy sheath
Bitumen
Polyethylene outer jacket
Semiconducting PE skin

Figure 5.3 Placement of an optical fiber in a ±400 kV DC 1050 MW XLPE insulated land cable. (*Source:* courtesy of Sumitomo Electric Industries.)

optical fiber in a separate duct (provided that duct is available for the fiber to be "blown into" it). It is to be noted that, under such circumstances, the sensing optical fiber is farther away from the cable conductor; hence, its response to the changes in cable conductor temperatures will not be rapid as compared to when it is placed under the cable sheath or jacket. If the optical sensing fiber

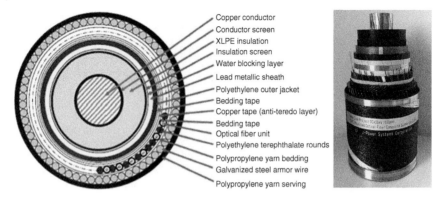

Copper conductor
Conductor screen
XLPE insulation
Insulation screen
Water blocking layer
Lead metallic sheath
Polyethylene outer jacket
Bedding tape
Copper tape (anti-teredo layer)
Bedding tape
Optical fiber unit
Polyethylene terephthalate rounds
Polypropylene yarn bedding
Galvanized steel armor wire
Polypropylene yarn serving

Figure 5.4 Placement of an optical fiber in a ±400 kV DC, XLPE insulated, 1050 MW submarine cable. (*Source:* courtesy of Sumitomo Electric Industries.)

Figure 5.5 Three-core distribution submarine cable with multiple fiber optic sensors. (*Source:* courtesy of Taihan Corporation, Korea.)

is farther away from the power cable, it may be only recording ambient soil temperatures! In certain duct bank constructions, it is quite common for the end user to select a telecommunication duct to place a temperature sensing fiber alongside the telecom cable to arrange for temperature monitoring.

The placement of the optical fiber has a direct impact on the interpretation of the power cable conductor temperature from the DTS measurements. The farther the optical fiber is placed relatively to the conductor, the more

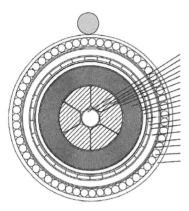

Figure 5.6 Schematic of a DTS sensing cable placed on a submarine cable and the physical arrangement. (*Source:* courtesy of BC Hydro.)

complex will the relationship be between the measured and computed conductor temperatures.

The accuracy of the relationship between the DTS and conductor temperatures is critical in helping the operator with the optimization of asset utilization and hence the power rating of the cable system.

Some examples of alternative fiber installations are shown in Figure 5.7.

The last diagram in Figure 5.7 shows an example where the electrical utility who has a high-pressure fluid-filled cable and also a spare steel duct has used the steel pipe as a conduit for placing the optical fiber for temperature monitoring. Typically, in such installation, it would be prudent to consider using a fiber in metal tube (FIMT) as a specific construction requirement for the DTS sensing fiber system.

Another point to be noted is that the thermal response of an optical fiber depends on the construction of the FO cable. This is because the multiple layers of insulation (thermal blanket effect) of the FO sensor slows down the thermal/time response. This could be significant in case of large bundles of fibers.

5.1.4 Example of Construction of the Stainless Steel Sheathed Fiber Optic Cable

Figure 5.8 shows a close-up view of three optical fibers placed into a 1.8 mm diameter stainless steel tube. The optical fibers are color coded for ease of recognition. However, if there is another set of stainless steel sheathed tubes in the cable geometry with the same fiber colors, this will prove to be a challenge during the splicing operation in a manhole. Unlike conventional power cables, there are no specific color coding rules that are applied in this industry and it is mindfulness of the utility engineer to ensure that they require the manufacturer to supply appropriate color-coded optical fibers to avoid erroneous splicing.

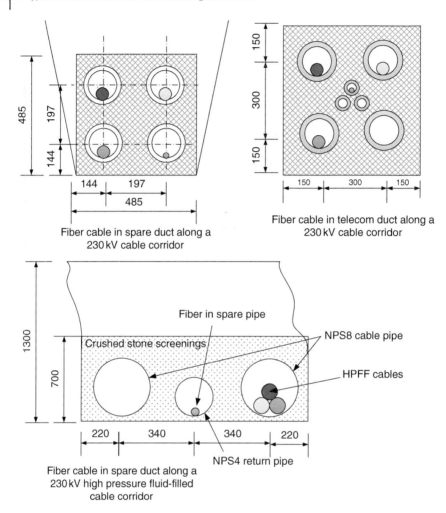

Fiber cable in spare duct along a
230 kV cable corridor

Fiber cable in telecom duct along a
230 kV cable corridor

Fiber cable in spare duct along a
230 kV high pressure fluid-filled
cable corridor

Figure 5.7 Schematic representation of different ways of installing DTS sensing fibers in
cable corridors.

5.1.5 Example of a Retrofit Placement of an Optical Cable into 525 kV Submarine SCFF Power Cable Conductor

5.1.5.1 Objectives of the Project

In 1995, with the support from BC Hydro's Strategic R&D Program and collaboration with the Electric Power Research Institute, a project was completed to demonstrate distributed fiber optic temperature sensing (DTS) applications to identify thermal bottlenecks on existing underground transmission cables. An EPRI report (Cherukupalli and MacPhail 2002; Cherukupalli et al. 1996)

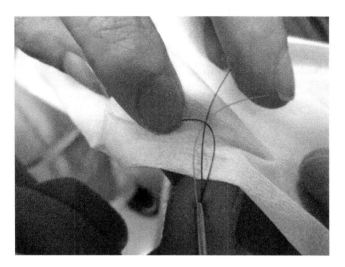

Figure 5.8 Close-up view of the optical fibers in a stainless tube. (*Source:* courtesy BC Hydro.)

was prepared which described a time progression of DTS applications at BC Hydro. This work also described the main benefits of using the technology such as increased high-voltage cable transmission capacity or avoidance of potentially damaging overloads. BC Hydro contracted a cable supplier to confirm the feasibility of inserting a 1.4 mm diameter stainless steel tube containing optical fibers into the conductor fluid duct in the actual submarine cables at one of the 525 kV substations (NCT). The prototype system was carefully evaluated for performance with high hydraulic pressures (2700 kPa) and temperatures (−26 °C to +82 °C) and its feasibility was then confirmed at a supplier's test facility. Following this successful research component of the investigation, a capital project was implemented. The main tasks were to insert optical fibers in each circuit at two terminals stations. In July 2003, optical fibers were placed into the conductor core of three cables (G1–G3) at the NCT substation. Subsequently, in July 2004, fibers were placed in the remaining cables (G4–G6) at the NCT substation and all six cables (M1–M6) at second substation (CCB) at the other end of the cable route.

The principal objectives of the overall project were to:

- Provide real-time conductor temperature monitoring and to allow the system operators to control the system in an optimized fashion.
- Study the limiting shore-end thermal characteristics of two submarine cable circuits to determine a more precise thermal rating of these circuits. This included the investigation of the effectiveness of the shore-end forced cooling system.
- Modify the shore-end cooling system and controls, as necessary, to optimize cable utilization and performance.

Two open-ended multimode fibers encased in a 1.8 mm stainless steel tube (FIMT cable) were inserted with suitable fluid seal fittings into the hollow cable core of the cables. Considerable effort and precautions were taken to control the hydraulic system pressure from the remote pressurizing plants to minimize the fluid flow during fiber insertion, without jeopardizing the integrity of the hydraulic system. Twelve 525 kV optical down links containing four multimode fibers were attached to the 525 kV bus and optical fibers appropriately spliced. Electric field analyses were performed to help design a special 525 kV_{rms} high-voltage corona-free optical cable splice case. This allowed the splicing of the FIMT cable inserted in the conductor core to the optical cable in the downlink while maintaining the splice junction at high voltage and adequately protected. The optical cable from the base of the downlink was then routed to the control building using suitable ducts and cable chases. The control building ends were pigtailed with special connectors to interface to the distributed temperature sensing (DTS) systems placed in the control rooms at NCT and CCB terminal stations, respectively. Figure 5.9 shows photographs of the field installation and the equipment in the control room at the NCT station. Both systems were capable of single-ended DTS measurements.

In all six cables, 0.5 m diameter loops were placed in the corona shields of the cable terminations to allow temperature calibration and confirmation of the measured values by the DTS unit. The length of these FIMT loops in cables G1–G3 and G4–G6 were 15 and 30 m, respectively. These loops provide a reference point to continuously confirm the accuracy of the temperature measurements outside and inside the power cable.

The DTS system was calibrated by placing this loop in hot and ice water to create reference isotherms and to confirm temperature measurements with the DTS unit. The surface temperature of the FIMT cable was recorded using a thermocouple and compared with the DTS system. Figure 5.9c shows a photograph of this site calibration process. The DTS unit was previously calibrated at the factory, but some tests were repeated to confirm data accuracy, reliability, and stability at the sites.

5.1.5.2 Installation

When the optical cables are installed externally to the power cable, they arrive at the site in the predetermined cut lengths. These lengths are governed by the distances over which this particular optical cable can be installed (such as Manhole-to-Manhole [MH–MH] distance or optical cable construction limitation). To complete the optical path, fusion splicing techniques will be required. This operation, if undertaken under suboptimal environmental conditions (outdoor environments, MH, etc.), can impact the quality of the splicing and hence affect the accuracy of the temperature measurement.

It is also important to keep in mind that the splices have to be mechanically protected from damage during installation and operation (splice case/box).

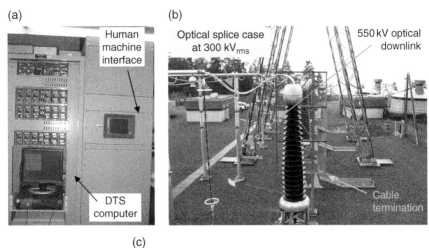

Figure 5.9 (a) Photograph showing the DTS unit, its computer and the human machine interface (HMI) panel that displays currents, fluid, armor, and air temperatures. (b) Photograph showing a panoramic view of the cable terminations at the NCT station with the optical downlinks connected to the bus. (c) Stainless steel-encased optical fiber in the plastic tube with ice water and hot water. (*Source:* courtesy of BC Hydro.)

Optical fiber splice box

Figure 5.10 Optical fiber splice. (*Source:* courtesy of BC Hydro.)

Considerations have to be given to ensure the nature/degree of the protection required from an environmental standpoint for this box (IP requirement). It is well known that if an unprotected fiber splice is exposed to water ingress, this can potentially lead to damaging the fiber itself and result in a premature failure.

When power cables have embedded optical fibers, the fibers have to be carefully trained out of the power cable and joint casing without violating the minimum bend radii to minimize and eliminate damage during extraction, assembly, and placement in the final resting position. The splicing procedure is more complex and depends on the power cable construction, grounding scheme, and so on. With the presence of the stainless steel tube, there is a likelihood of an electrical hazard due to electrical induction. Extra attention has to be paid in such a case to ensure that the exit points of the optical fiber from the power cable as well as the splices are completely watertight and mechanically and electrically protected in order to avoid accidental damage during installation and operation. Any damage to the FO element can have a direct and serious impact on the integrity of the power cable system itself (power cable failure due to water ingress, short circuit, etc.). Once the fibers are extracted, they have to be precisely spliced to ensure that the optical loss across this connection is well within the prescribed limits.

Figure 5.10 shows an example of the optical fibers exiting the left and right ends of the joint through the joint casing and being brought together into a splicing tray and once completed, placed into a sealed casing and held tightly against the cable joint casing by nylon tie-down straps.

Figure 5.11 shows another example of a cable with an optical splice case that will be directly buried.

Figure 5.11 A fiber optic joint box location relative to an XLPE cable joint. (*Source:* courtesy of Nexans.)

One of the practical challenges with embedding and splicing of optical fibers in power cables is that generally this is a specialized work and needs good integration between the cable men who are proficient with the handling and splicing of power cables with the FO splicers who have no knowledge of the power cables but are experienced in handling optical fibers. These two teams have to work in close cooperation in a confined space (as most of this work is done in manholes in North America). Further, the roles and responsibilities of these two crews and how and when they interact have to be coordinated with great precision to achieve success. Sloppy workmanship with optical splicing can lead to challenges down the road.

The presence of a splice case also provides a unique opportunity to individually heat each casings in each manhole and monitor the casing temperature accurately using classical thermocouples or RTD and compare these artificially created "hot spots" with the DTS system (see Figure 5.12). This provides two benefits, namely:

- The ability to locate and isolate the manhole on a temperature versus time trace the DTS produces.
- The ability to isolate any hot spots along the cable route and determine their approximate geographical distances from the identified manholes.

5.2 Advantages and Disadvantages of Different Placement of Optical Fibers in the Cable

Let us consider a situation where the optical fiber is not too close to the conductor. In such a case, sudden changes to the cable conductor temperature due to an increase in the load current may not be recorded by the optical fiber placed farther away from the cables. On the other hand, if the optical fiber is integrated with the power cable, the temperature changes experienced by the cable will also be recorded faithfully by the optical fiber. This proximity of the sensing optical fiber and its relative closeness to the cable conductor will become quite vital when the utility planning to use the DTS data in a real-time thermal rating (RTTR) software. Clearly, the closer the optical fiber is to the cable conductor, the better the accuracy of the prediction by the RTTR software will be.

A concern has been expressed with the embedment (integration) of an optical fiber inside the core of a power cable. The question that has been raised is that while the design life for a power cable is usually considered to be 40 years, the life of an optical fiber is not really well known. Unlike in the case of the telecommunication industry where the optical fiber is only subjected to optical switching, when such fiber is embedded in a power cable, it is also exposed to severe thermomechanical bending (TMB) stresses due to load cycling during

(a)

(b)

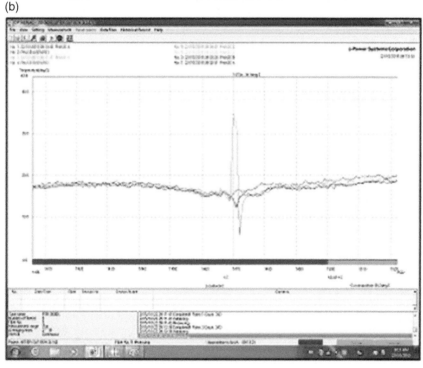

Figure 5.12 (a) Heating of the splice case with a heater and (b) recorded DTS trace showing the temperature spike and location. (*Source:* courtesy of BC Hydro.)

the cable operation. Would such TMB harm the optical cable and preclude the possibility of it yielding accurate temperature data over the same life cycle period? In other words, will the optical fiber also yield a 40 year life? Should the embedded fiber physically get damaged due to TMB stresses, there is no opportunity to undertake DTS measurements. On the other hand, if the optical sensing fiber is external to the power cable no such shortcomings exist. It is some of these concerns that have led to decisions by some utilities to place the optical fiber outside of the power cable.

5.2.1 An Example with Placement of FO Sensors at Different Locations Within the Cable Installation

This real-life example from the BC Hydro systems illustrates the effect of placing the fiber optic sensors at two different locations within the same cable and outside. Figure 5.13 shows a drawing of a 525 kV submarine cable with two layers of copper armor and a protective bedding layer buried inside a concrete cable chase with water inlet and outlet pipe placed beside and above the power cable. Figure 5.14 shows a close-up side view of the actual field assembly with the cable chase exposed to apply the RTD on the pipes and the cable.

For the purpose of redundancy, two platinum RTDs were placed at the locations described above.

Figure 5.15 shows an example of real-time measurements from 31 May to 7 June, approximately 39-day period, during which the DTS system was recording the real-time conductor temperature and the RTDs placed on the cable were monitoring the cable surface temperature. The load current and the coolant

Figure 5.13 Schematic of the cable placed beside water cooling pipes with temperature monitored inside the hollow conductor, on the surface of the cable layer, and estimated on the surface of the concrete chase.

Figure 5.14 Close-up view of a single-phase 525 kV cable buried inside a cable chase with forced cooling pipes and RTD applied on the cooling pipes and on the power cable to monitor surface temperatures. (*Source:* courtesy BC Hydro.)

flow rates were also monitored during the same period. A simple comparative analysis of the data shows the following:

- The difference between the conductor and the cable surface temperatures is approximately 12.75 °C for the prevailing load conditions.
- Once the cooling system is shut down, the cable conductor and surface temperatures tend to increase. There is some variability in the armor temperatures and the difference between the conductor and the cable's surface temperatures tends to increase as seen with the downward trend in the graph.

These comments refer to the prevailing loads. It remains uncertain what happens if the load now suddenly doubles. The relationship between the DTS conductor temperature and the cable surface temperature may be more divergent.

The relationship between the temperature measured by the DTS system on a fiber and the actual conductor temperature is complex. This becomes essentially more complex farther the optical fiber is placed from the cable conductor as the measured temperature will be influenced by the environment it is immersed into.

The experiment has demonstrated that when the DTS fiber is farther away from the power cable, the ability to predict the conductor temperature becomes difficult as the intervening materials with unknown (or temperature-dependent) thermal resistivities will influence the inference (calculation).

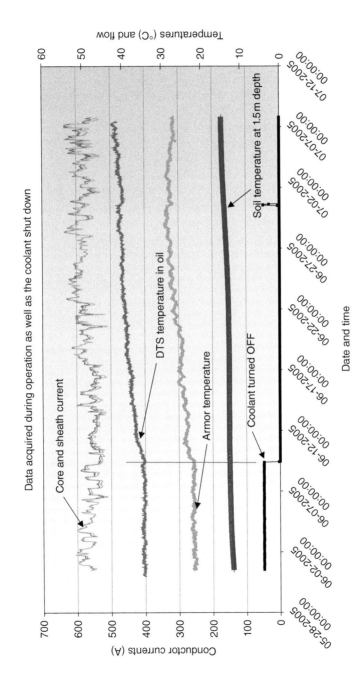

Figure 5.15 Measured current and temperatures in a 525 kV cable.

5.3 What Are Some of the Manufacturing Challenges?

When optical fibers are placed between copper wire neutrals as a means of carrying the required fault current, it is relatively easy to incorporate a stainless steel encased fiber optic cable by replacing one of the copper wire neutrals. For example, during the winding operation for cable production, one will need to replace the normal copper wire bobbins with bobbin containing the stainless steel sheathed fiber optic cable. The tension during the winding operation will have to be controlled carefully to ensure that no kinking of this stainless tube can occur since this can potentially damage the optical fiber. Adjustments will have to be made during the operation should the cable conductor or wire diameters change.

In contrast, if the stainless steel sheathed optical fiber has to be placed in an XLPE cable with an extruded smooth or corrugated sheath, the challenges become more severe. The risks of damage to the fiber increases as there are two additional mechanical stressful operations that have to be performed; namely, extrusion of the aluminum sheath over the cable core and then subjecting it to corrugations. Figure 5.16 shows an example in which the stainless steel sheathed cable was surrounded by two additional copper tubes to mechanically protect the assembly. Another layer of semiconducting tape applied helically provided improved anchoring of the optical fiber to the cable core before the aluminum extrusion operation was commenced. This introduces additional production time and can add to the cost of the cable.

Some high-voltage cable sheath designs use an aluminum sheet typically 1.8–2 mm thick which is wrapped and tightened around the cable and the seam

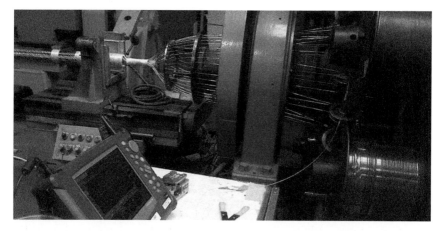

Figure 5.16 Applying the copper wires and the stainless steel sheathed cable. (*Source:* photo courtesy of SEI.)

is laser welded. The ability to embed an optical fiber in such a cable construction is almost impossible. There are concerns about potentially inflicting damage to the optical fiber during the "squeezing operation" or even during the laser welding. Manufacturers that produce such cables with welded sheath designs use an external optical fiber system.

While there are significant advantages in having the DTS sensing fiber closer to the conductor as in an embedded design for the reasons described in Section 5.2.1, such designs also offer another set of challenges; for example, how can one safely extract an optical fiber from the core without causing cable damage. This is a very important consideration and the users should actively engage the cable manufacturers to learn about these issues and seek appropriate information to improve the confidence that such fiber extraction are done and premature failure is avoided.

6

DTS Systems

This chapter contains an overview of the architecture of a DTS system and provides information on the requirements of quality assurance and factory acceptance tests as well as nature of the reporting that ought to be followed in the context of a utility environment. It also addresses the impact of the sensitivity and inaccuracy of a DTS system.

6.1   What Constitutes a DTS System?

DTS systems typically have the following five major hardware components:

- Optical fiber
- Laser and focusing systems.
- Photodetector and optical-electrical processing unit.
- Controller (data processor).
- Optical multiplexing switch (sometimes built into the DTS system or may be external to the system) allowing temperature measurements along several different fibers in a sequential manner.

In addition, it also has communication software that allows two-way communication between an external computer as well as a supervisory control and data acquisition (SCADA) system to transmit information to the control center of a utility. The exact architecture of a DTS system varies considerably between different providers. Some have an inbuilt computer processor that will allow automatic control, measurements, and storage whereas some other may require an external computer to perform these functions.

A schematic description of a typical DTS system using Raman and optical time domain reflectometry (OTDR) is shown in Figure 3.6 and repeated in Figure 6.1.

Figure 6.2 shows the schematic of a DTS system integrated into the control center of a power utility but at the same time allowing the engineers to access

Distributed Fiber Sensing and Dynamic Rating of Power Cables, First Edition.
Sudhakar Cherukupalli and George J. Anders.
© 2020 by The Institute of Electrical and Electronics Engineers, Inc.
Published 2020 by John Wiley & Sons, Inc.

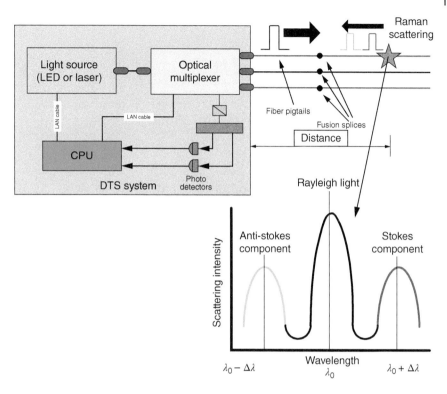

Figure 6.1 Typical DTS system using Raman and OTDR technology. (*Source:* BC Hydro.)

the DTS computer remotely for any adjustments and troubleshooting using a separate telecom network.

6.2 Interpretation and Application of the Results Displayed by a DTS System

6.2.1 General

Today, the zoning and alarming features of most DTS software, can prove useful for alerting the operators to abnormal conditions. While operators generally do not care too much about the location of the hot spot but are more interested in the resulting power transfer limit imposed on the network from this alarm warning. However, at a more basic level, the detailed temperature profiles reveal a wealth of information to cable engineers about the ambient, time-varying thermal environment that the cables are installed in. By "thermal environment" we mean the thermal characteristics of materials surrounding

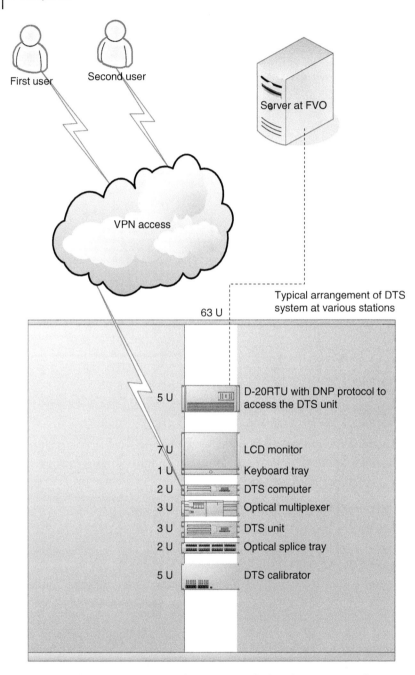

First user

Second user

Server at FVO

VPN access

63 U

Typical arrangement of DTS system at various stations

5 U — D-20RTU with DNP protocol to access the DTS unit

7 U — LCD monitor
1 U — Keyboard tray
2 U — DTS computer
3 U — Optical multiplexer
3 U — DTS unit
2 U — Optical splice tray

5 U — DTS calibrator

Figure 6.2 A schematic architecture of a DTS system deployed at a station and reporting data to the system control center at the same time allowing VPN-type access to the DTS computer. (*Source:* BC Hydro.)

the cables. For example, for cables in air, ambient temperature, wind speed, and solar radiation are important parameters. For buried cables, ambient soil temperature is very influential and varies seasonally and with depth. Solar heating onto asphalt surfaces can also influence ambient soil temperatures. Soil thermal resistivity and its thermal capacitance, which are heavily influenced by moisture content, are also important factors, varying with rainfall and the presence of vegetation in close proximity to cables.

The temperature rise experienced by loaded cables is the ambient temperature plus cable temperature rise caused by voltage and current. Since the cable temperature rise can be calculated and understood much more clearly than that of an axially nonhomogeneous thermal environment, in a broad sense, DTS systems help to improve understanding of the actual thermal environment experienced by the cables, more than understanding about the cables themselves. With this in mind, it is often of high value to check DTS temperature profiles shortly before initial commissioning, or during long circuit outages, to "benchmark" ambient temperatures and to locate hotspots caused perhaps by other utilities, such as steam heat lines, distribution duct banks, and so on. However, this will not help to determine soil thermal resistivities and capacitances along the route. For that, careful desktop study and analysis of cable temperature response under load is needed. The exception would be if a DRS with DTS temperature feedback was applied.

6.2.2 Comparison of Measured and Calculated Temperatures

This section describes a case study illustrating how DTS profile information can be applied to infer characteristics of circuit thermal environment and to make adjustments in power cable loading or improvements to a poor thermal environment.

Figure 6.3 shows a cross-section of a 230 kV SCFF cable with 1250 kcmil hollow copper conductor installed in a 2 × 2 duct bank.

This 7.5 km-long cable system is placed in 2 × 2 concrete duct bank located in a suburban area of greater Vancouver. Along a stretch of approximately 300 m, the duct bank runs along a boulevard and there are some tall trees along this corridor. Almost 30 years into its service, a condition assessment report made a recommendation that an optical fiber be blown into the spare duct and integrated with a DTS system to measure the cable temperature along this corridor.

When reviewing the DTS data from a measurement in a spare duct, it became obvious that despite the relatively light loading conditions, the average duct temperature hovered around 27.5 °C. The DTS traces from measurements along this route displayed some anomalous temperature features where the temperature maximum varied between 32.5 and 36 °C, suggesting that these could become thermal bottlenecks in the future and restrict the ability for the

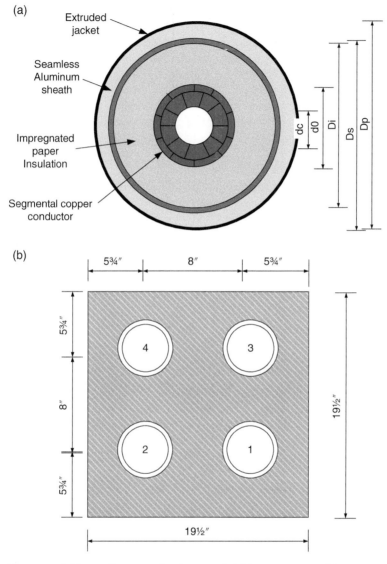

Figure 6.3 Cable installation used in the test. (a) Cable cross-section; (b) cross-section of the 230 kV duct bank.

cable system to deliver the rated power through this corridor. An attempt was made to study the cause of these anomalous features by selecting temperature readings at Zone 2 on Channel 5. This zone was defined to be the region between 4860 and 5000 m along the FO cable corridor.

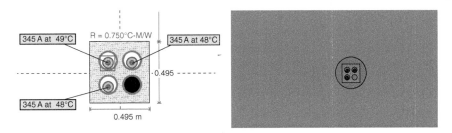

Figure 6.4 CYMCAP and FE model of the 230 kV SCFF configuration. In the model, a zone near around the duct bank shows soil drying due to large pine trees over the duct bank.

A pre-existing CYMCAP model was refined to match the DTS measurement of the FO cable in the spare duct with the calculated temperature. The key parameter altered was the soil thermal resistivity close to the duct bank. Figure 6.4 shows the model representing the duct bank at Zone 2. Assuming the soil thermal resistivity of 1.2 °C-m/W, a discrepancy of 2.5 °C was observed between the measured DTS and calculated spare duct temperatures. This analysis was conducted with data from 31 October 2014 at 11:00 a.m. when the cable was carrying a load current of 345 A. By increasing the assumed soil thermal resistivity in the vicinity of the duct bank, a better match of the DTS and calculated fiber temperatures was achieved. To understand the cause, a route patrol was conducted along the corridor and it became clear that the location where anomalous temperature features were observed on the DTS traces coincided with the presence of large pine trees growing near the duct bank. It was strongly suggested that a local change in soil thermal properties may be the cause, as the tree roots absorb moisture in the ground leading to soil dehydration, higher thermal resistivity, and thus, higher observed temperatures.

This analysis was only undertaken to illustrate the possible reasons for higher temperature zones appearing along the route based on the recorded DTS measurements.

Armed with such information, the real-time thermal[1] capacity (MVA) of the cable corridor can be developed and used for the real-time rating analysis. Today, many DTS suppliers incorporate rating calculation software into

---

1 The technical literature frequently refers to "real-time" implying instantaneous measurements and calculations. In practice, however, the DTS measurements can sometimes approach several tens of minutes so application of real-time ratings need to take such time delays into account. The measurement time is governed by the desired accuracy (hence, averages of repeated measurements) and length of fiber being monitored.

their system or into the local DTS computer. The user can initially select the thermally limiting section and then the rating calculation software can provide the maximum thermal capacity of the cable system in real time using the instantaneous DTS "hot spot" temperature. In this manner, the system control operators will be equipped with the real-time rating information and will allow them to operate the system in an optimal manner. This topic is addressed in more detail in Chapter 9.

6.3 DTS System Calibrators

It is very important to remember that the recorded DTS temperature is dependent on several parameters such as the laser power, optical characteristics of the sensing fiber, the long-term behavior of the connectors, and fusion splices along the sensing corridor; all of which can affect the overall system accuracy. If system operators are relying on this measurement to optimize asset utilization, the reliability and accuracy of the temperature measurement becomes very important. Therefore, it would be prudent to adopt annual DTS system calibrations as a routine maintenance practice. One of the ways to simplify this exercise is to incorporate a DTS calibrator at the end of the monitoring corridor, so that when placed in series with the optical path it will show two "preset and known" temperatures.

The DTS system calibrators are discussed in detail in Chapter 7.

6.4 Computers

While DTS systems have evolved and have been extensively used in the oil and gas industry, their use is increasing rapidly in the electrical utilities. Computers and other auxiliary hardware used to control the DTS system are subject to severe transients in the power system applications, making them prone to damage. Temperature excursions can also cause computers to malfunction. Therefore, it is important to consider the test and performance requirements of these computers and optical multiplexers in such a harsh environment. Ensuring that the electronics use the DC power is one solution option. The utility engineer should discuss with the supplier how these aspects of the DTS system are being verified for performance under arduous high-voltage power system conditions. Electromagnetic compatibility (EMC) compliance requirements will be one way to ensure the long-term performance of the computers connected to the DTS system. These requirements, described in Section 6.5, model specification for DTS systems.

In certain designs, the DTS system components and control systems are heavily integrated into one box, resulting in potential thermal-related stability issues. For example, some DTS suppliers have a 30 °C upper temperature limit for their equipment, which can be exceeded in some operating locations. Care has to be exercised in the specifications that under these operating conditions, the DTS system will continue to provide reliable, stable, and accurate data. A complete system "burn-in" test done during the FAT is suggested in agreement with the DTS systems supplier.

6.5 DTS System General Requirements

Section 6.5.1 briefly describes the general requirements for a DTS system based on the experience at a specific utility.

6.5.1 General Requirements

The general requirement can be specified as follows: "The Contractor shall supply a Distributed Temperature Sensing (DTS) system that will be used to monitor the temperature continuously along the specified circuit(s). The system shall perform by detecting backscattered laser light, calculating the ratio of the Raman Anti-Stokes to Stokes signal intensities, and determining temperature measurements incrementally along the entire length of the fiber using computer controlled time-of-flight measurements."

More specifically, the system shall include but not be limited to the supply of the following components:

- regulated and self-contained DC power supply
- laser launch and detect electronics
- optical multiplexing units
- software control to perform various access and logging functions
- software to control the launch optics/focusing
- control hardware for performing storage, monitoring, and alarm functions
- front or rear patch panels to connect the sensing fibers and their corresponding pigtails
- computer/dial-in/network hardware to connect and access the DTS unit locally as well as remotely

Typically, the DTS unit will be installed at a substation and will be interfaced with a remote terminal unit (RTU) to provide temperature data to the control center, at user-defined intervals set by the control computer. In addition, the DTS unit should be capable of remote access by a dial-up/IP configurable connection.

6.5.2   Summary of Performance and Operating Requirements

Some of the performance and operating requirements are summarized below. These are based on an actual cable installation in a North American utility and should be adjusted to match the local conditions.

1) DTS instrument shall be capable of measuring temperature on at least six channels (single-ended measurements) with a multimode communications grade 50/125 μm (core/cladding) sensing optical fiber over a minimum length of 8 km of the sensing fiber, or at least half the total power cable route length, such that complete coverage of the cable route can be achieved. These measuring distances shall be determined from the instrument to the remote end of the sensing fiber, which in the case of a closed-loop system will be near the instrument.

2) If, for operational reasons, there is a need to use closed-loop sensing fibers, the DTS system shall be suitable for use with a minimum of six closed-loop sensing fiber configurations, each with a total sensor capability of 8 km, or at least half the total power cable route length.

3) The DTS system shall be capable of obtaining the temperature data from each channel in either closed-loop or open-loop configurations with the basic accuracies specified.

4) The basic spatial resolution for measurement using open-ended sensing fibers shall be 0.5–1.5 m and the basic temperature resolution shall be 0.5–1.5 °C, over at least half the required power cable route length.

5) The DTS instrument shall be optimized for use with 50/125 μm multimode graded index optical fiber cables meeting the requirements of latest version of ITU Publication G.651.1 (2016).

6) The DTS instrument shall have suitable bulkhead connectors to which the suitably ruggedized pigtail fibers can be connected. Some DTS manufacturers prefer a pigtail with a bare E2000/UPC 0° butt-cleaved or E2000/APC 8° angle-polished end, which can be fusion spliced to the system fiber to achieve the stated accuracy and resolution. But this is generally not the customer responsibility and must be part of the equipment design. The bare ends of the pigtail fibers can be connected to the sensing fibers in the following manner, without compromising the basic measurement accuracy and resolutions as specified herein, or otherwise provided by the supplier:
   a) mechanically spliced
   b) connected
   c) fusion spliced.

7) The instrument bulkhead connector shall be interchangeable to accommodate various types of connectors such as FC, SC, E-2000, and ST.

8) The operating lifetime for the DTS system shall be a minimum of 10 years continuous operation. During this period, assuming that the optical fiber has not degraded, the system accuracy and resolution shall be preserved.

The DTS system shall have a minimum mean time between failures of not less than 50 000 hours. If an optical switch is used, the mean time between failures shall be greater than 10 million switch cycles.

9) The system accuracy and resolution shall not be compromised for at least 100 repeated connect and disconnect operations of the pigtail fibers for a specified optical attenuation variation of between 0.0 and 0.4 dB or as deemed appropriate by the manufacturer.

10) The laser output of the DTS unit shall meet the Class I Laser Safety requirements.

11) The DTS system when used in the OTDR mode shall have the following performance specifications:
    a) With a signal/noise ratio (SNR) = 10.
    b) Return loss better than 54 dB for two seconds averaging.
    c) Return loss better than 74 dB for two minutes averaging.
    d) With a SNR = 20 single-point resolution should be 1 m.
    e) With a SNR = 5 single-point resolution should be better than 5 m.

12) DTS system should be operable with any of the following power supply specifications:
    a) 120 V/240 V ± 10%, 50 Hz/60 Hz ± 5% single-phase power supply (preferred).
    b) 24 V/48 V/125 V DC supplies.

13) The DTS system shall comply with the International Safe Transit Association (ISTA) Impact Test – Procedure 1A Drop Method and ISTA 1 Vibration Test – Method B (1.1 g) or equivalent.

14) The DTS system and display equipment shall be rack mounted in terminal station relay rooms with at least one end of the power cable circuit, depending on power cable route length and DTS system range.

6.5.3 Electromagnetic Compatibility Performance Requirements for the Control PC and the DTS Unit

The DTS instrument along with the computer should be capable of working without compromising the accuracy and resolution specification in substation environments where switching transients and radio frequency (RF) interference are present. The supplied equipment should meet the basic EMC requirements for radiated interference as well as mains-borne interference. Power and control cables should be shielded to reduce induced overvoltages when used at site. The computer and test equipment should be compliant with the latest versions of the following standards:

1) IEC 61000-4-1 Electromagnetic Compatibility (EMC) – Part 4–1: Testing and Measurement Techniques – Overview of IEC 61000–4 Series.

2) IEC 61000-4-2 Electromagnetic Compatibility (EMC) – Part 4–2: Testing and Measurement Techniques – Electrostatic Discharge Immunity Test.

3) IEC 61000-4-3 Electromagnetic Compatibility (EMC) – Part 4–3: Testing and Measurement Techniques – Radiated, Radio-Frequency, Electromagnetic Field Immunity Test.

4) IEC 61000-4-4 Electromagnetic Compatibility (EMC) – Part 4–4: Testing and Measurement Techniques – Electrical Fast Transient/Burst Immunity Test.

5) IEC 61000-4-5 Electromagnetic Compatibility (EMC) – Part 4–5: Testing and Measurement Techniques – Surge Immunity Test.

6) IEC 61000-4-6 Electromagnetic Compatibility (EMC) – Part 4–6: Testing and Measurement Techniques – Immunity to Conducted Disturbances, Induced by Radio Frequency Fields.

7) IEC 61000-4-8 Electromagnetic Compatibility (EMC) – Part 4–8: Testing and Measurement Techniques – Power Frequency Magnetic Field Immunity Test.

8) IEC 61000-4-9 Electromagnetic Compatibility (EMC) – Part 4–9: Testing and Measurement Techniques – Pulse Magnetic Field Immunity Test.

9) IEC 61000-4-10 Amendment 1 – Electromagnetic Compatibility (EMC) – Part 4: Testing and Measurement Techniques – Section 10: Damped Oscillatory Magnetic Field Immunity Test. Basic EMC Publication.

10) IEC 61000-4-11 Electromagnetic Compatibility (EMC) – Part 4–11: Testing and Measurement Techniques – Voltage Dips, Short Interruptions and Voltage Variations Immunity Tests.

11) IEC 61000-4-12 Electromagnetic Compatibility (EMC) – Part 4–12: Testing and Measurement Techniques – Ring Wave Immunity Test.

12) IEC 61000-4-29 Electromagnetic Compatibility (EMC) – Part 4–29: Testing and Measurement Techniques – Voltage Dips, Short Interruptions and Voltage Variations on DC Input Power Port Immunity Tests.

13) CISPR 11: Limits and Methods of Measurement of Radio Disturbance Characteristics of Industrial, Scientific and Medical (ISM) Radio Frequency Equipment.

14) ANSI/IEEE Standard C37.90.1-1989 IEEE Standard Surge Withstand Capability (SWC) Tests for Protective Relays and Relay Systems.

15) IEEE C37.90.2 Standard for Withstand Capability of Relay Systems to Radiated Electromagnetic Interference from Transceivers.

16) IEEE/ANSI C62.41.1-2002 Guide on the Surge Environment in Low-Voltage AC Power Circuits.

17) IEEE/ANSI C62.41.2-2002 Recommended Practice on Characterization of Surges in Low-Voltage AC Power Circuits.

6.5.4 Software Requirements for the DTS Control

This section outlines the software requirements for the DTS control and is based on a practical guide used in one of the North American utilities.

1) Supplied software shall be a Microsoft Windows-based program with menu-driven capabilities as well as graphical user interface (GUI) for facilitating the measurement setup, including: cable lengths, zones, trace ID, and alarms.

2) Generated data file structure format should be compatible with Microsoft Excel or Microsoft Access spreadsheet programs to allow for dynamic object linking with other Windows-based application programs.

3) Software shall permit easy user-labeling of zones (distance and mean, high or low temperatures in the zones), hot spot locations, trace annotations, and features alarms for each measurement.

4) Software shall have an Auto-Save feature that allows user-controlled, time-driven closing of data files as well as all other important parameters (defined by the Supplier) such as fiber attenuation characteristics and OTDR-related information, and so on.

5) Software shall permit periodic and automatic monitoring as well as recording of optical attenuation in sensing fibers at user-defined intervals.

6) Software shall be capable of allowing remote access and control of the DTS unit either via Ethernet or via telephone dial-in.

6.5.5 DTS System Documentation

The DTS supplier should provide a DTS System Report, describing the system to be supplied, including:

- Technical documentation stating DTS system power supply requirements, accuracy in temperature measurements, spatial resolution, maximum allowable fiber attenuation at operating frequencies, and measuring intervals, for various fiber ranges.
- Descriptions and interface requirements with the terminal station SCADA systems and DRS system, if specified.
- Descriptions of output protocols to SCADA for eventual display of real-time temperature profiles and zone alarm data to remote system operators.
- Descriptions of methods to be used by remote specialists possibly from the vendor to modify DTS system settings, oversee functionality, and retrieve historical data.
- In addition, detailed documentation of the DTS system should be delivered as listed below:
  - Schematic overview of the system
  - Test reports (including FAT and SAT)
  - Material list
  - Electrical drawings
  - Lay-out drawings, including as-built drawings
  - User manuals
  - System failure warnings
  - Maintenance prescriptions and schedule

7

DTS System Calibrators

This chapter discusses the use of DTS calibrators and how they may benefit in improving the accuracy of the DTS data over the long term.

7.1 Why Is Calibration Needed?

It is very important to remember that the recorded DTS values using a Raman- or a Brillouin-based system for temperature measurements are dependent on several parameters such as:

- The stability and intensity of the laser power and its power density.
- The preservation of the optical characteristics of the sensing fiber.
- The long-term behavior of the connectors and fusion splices along the sensing corridor.
- The reliability of the optical detectors and filters.
- The stability of the various digitizers, averaging modules, frequency analyzers, and miscellaneous electronic components.
- The software and the computer that controls the DTS system.

All the above parameters can affect the overall system accuracy.

Such systems have to monitor cable circuits with an expected design life of 40 years. If the DTS users have to provide accurate temperature data, especially under system overloading conditions, to avoid damage to the expensive cable asset, the stability, accuracy, and spatial resolutions of the measurements are very important.

Quite often, these temperature-measuring fibers are either attached to or embedded in power cables and can experience thermomechanical stresses while in service. The long-term effects of these stresses on the optical cable's longevity remain unknown. Moreover, in the optical path between the fibers at each end of a joint or between the cable termination and the DTS, there are

Distributed Fiber Sensing and Dynamic Rating of Power Cables, First Edition.
Sudhakar Cherukupalli and George J. Anders.
© 2020 by The Institute of Electrical and Electronics Engineers, Inc.
Published 2020 by John Wiley & Sons, Inc.

many intervening fusion splices, connectors, and other possible items that are employed for various reasons. The long-term characteristics of these sites can also affect the DTS unit's thermal performance. The presence of any moisture can possibly also result in optical fiber power losses decreasing the available optical budget and affecting the accuracy of the reported temperature measurement. Therefore, it would be prudent to undertake some means of calibration of the optical fiber to ensure the reliability of the measured temperature at the commissioning stage as well as at periodic intervals.

An important consideration is how often should one undertake such calibration and how should they be done. Sections 7.2 and 7.3 address these questions in some detail.

7.2 How Should One Undertake the Calibration?

At present, there are no industry standards that prescribe either the method or the frequency of calibration of the DTS systems. A practice that has been adopted at BC Hydro is to undertake field calibration of a DTS system (which is referred to as the site acceptance tests) and then annually repeating this process. Figure 7.1 shows a typical DTS calibrator. This can be battery powered for periodic use or AC powered for continuous application. It essentially contains two sets of fibers that are placed in a carefully controlled temperature environment with input terminals to accept standard optical connectors, defined by the DTS system requirements. When energized, it will provide two user-settable reference temperatures such as, for example, 70 and 30 °C.

Figure 7.1 Photograph showing a typical DTS calibrator (black box) placed at the end of an optical path along with a fiber termination box. (*Source:* courtesy of BC Hydro.)

These DTS calibrators may be purchased from some of the DTS vendors. The units can be custom designed and the designs evolved from the collective experience based on the performance of some of the older units. To the DTS world particularly, in the oil and gas industry such a need for accurate and calibrated temperature measurements are less onerous but they become very important in the electric utility market.

Some older designs had problems such as:

1) The temperature controller did not appear to be working.
2) The fibers inside the DTS calibrators were very fragile and prone to damage when one opened the door to operate the circuit breaker and power up the unit.
3) There were no connectors and fusion splicing was required between the test fiber and the fiber in the DTS calibrator unit.
4) The front panel LEDs were broken.
5) The temperature measurements seemed to be performed by low-cost thermocouples and were not high-quality platinum RTDs that demonstrate much better stability and accuracy. These needed to be replaced.
6) The calibrators were purchased at a time when no manufacturer actually supplied suitable field-mountable units.

The challenge for the classical calibration is its costs. This is particularly affecting the DTS systems placed at remote control room locations that require a day of travel. During the calibration, a circuit outage may be required to satisfy safety requirements such as:

- Isolation of the circuit.
- Isolation of the strategically placed FO cable calibration loops (so they may be immersed in the hot and cold water baths).
- Undertaking the measurements.
- Removing and restoring the system to allow circuit re-energization.

Increasingly, getting an outage for such maintenance activities is a considerable challenge at many utilities.

As an alternative, BC Hydro pioneered a method to encourage DTS manufacturers to build in-line DTS calibrators. Section 7.3 describes the merits of such systems and describes in a diagrammatic form a simple actual application that has been built to provide continuous online calibration.

Figure 7.2 shows a schematic of a DTS calibrator placed at the end of the cable monitoring corridor in series with the optical path. The DTS calibrator may have one to two thermowells that have approximately 20–30 m of optical fiber which is maintained at user-settable temperatures. Some designs also incorporate a system whereby the user can use two "preset and known" temperatures, namely hot and cold thermowells. There is a section between the two thermowells where the optical fiber will be at the box's ambient temperature.

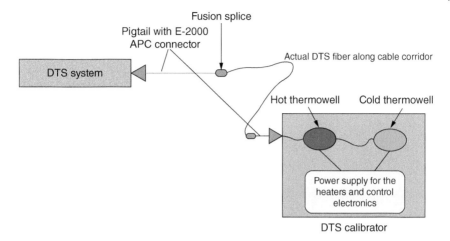

Figure 7.2 Schematic of a DTS calibrator.

These thermowells have to be accurately controlled to maintain the required temperature. When the DTS unit is connected to the measuring fiber optic cable as well as to the calibrator in series, the expectation is that the DTS system will not only record the FO cable temperature along the cable route but also measure the temperature of the fibers in two preset temperatures on the calibrator.

The control software built-in to operate the DTS should also have capability to recalibrate the DTS unit automatically when connected to the calibration unit.

7.3 Accuracy and Annual Maintenance and Its Impact on the Measurement Accuracy

Figure 7.3 shows a typical cable termination box and a DTS calibrator connected optically in series. The black unit on the right is the DTS calibrator. This unit can be connected to either AC or DC power. Ideally, if it is placed in a substation environment, it would be prudent to power the unit using DC power as such supplies are more robust and immune to the SF_6 GIS switch-induced transients.

A typical DTS calibrator shown in Figure 7.3 contains two sets of fibers that are placed in a carefully controlled temperature environment with input terminals to accept standard optical connectors, defined by the DTS system requirements. When energized, it will provide two user-settable reference temperatures, such as 70 and 30 °C.

Figure 7.3 DTS calibrator. (*Source:* courtesy of BC Hydro.)

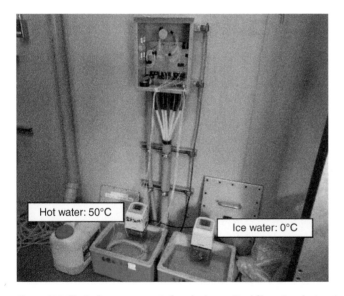

Figure 7.4 Typical arrangements for placing optical fibers in a hot water bath and ice as a means of calibrating the remote end. (*Source:* courtesy BC Hydro.)

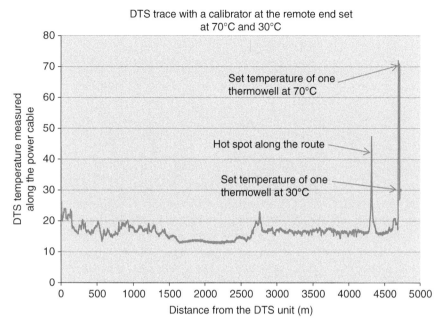

Figure 7.5 DTS trace showing the remote end of the cable connected to a DTS calibrator set at 70 and 30 °C.

Quite often it is the manufacturers' recommendation to undertake the DTS calibration by applying the thermal environment at the remote cable ends. The rationale is that when light is injected into the cable, optical attenuation occurs naturally. The longer the cable the higher is the attenuation. For example, for a Raman-based DTS system, the temperature relies on the anti-Stokes component of the reflected light, which is typically three orders of magnitude weaker than the injected light. Therefore, a certain optical budget is needed to allow meaningful and accurate measurements. By placing a remote fiber end at controlled temperature and confirming that the DTS unit measures this temperature accurately improves the confidence in measurements all along the cable.

Quite often, in land cables there are underground vaults where cable sections are joined. These are also locations where the optical fibers are also joined. It is also possible to undertake similar calibration in a manhole so that the entire route provides accurate temperature data. With time, should there be an increase in optical attenuation or some such anomaly, monitoring the remote end which is connected to a DTS calibrator, which has been set to two preset temperatures, will confirm whether there is any loss in accuracy. Figure 7.4 shows how the remote and accessible cable ends are placed in a hot water bath and ice for calibration purposes. Figure 7.5 shows a DTS trace with the approximately 3.5 km-long optical fiber connected to a DTS calibrator at the remote end. The calibrator in this case is set to two temperatures, namely 70 and 30 °C.

8

DTS System Factory and Site Acceptance Tests

A power cable is subjected to several tests during its life. It begins with a series of prototype tests on the materials as well as on a cable itself. It is then subjected to a year-long constant voltage and load cycling tests simulating its placement in different environments (such as in tunnels, joints placed in manholes, buried in duct banks, and so on). Since these tests are very long, time consuming, and expensive, the designs are also subject to type tests. Further, a sample of the manufactured batch is subjected to routine factory tests and finally once it is placed in the client's facility, subjected to a site acceptance test (SAT). During its operating life, it is also subjected to routine maintenance tests to ensure its longevity and reliable service life.

If one were to draw a parallel for assessing and ensuring the long-term performance of DTS system, it would seem logical to subject it to a similar series of tests. One important variant between the DTS system and the analogy drawn with the power cable is that quite often the DTS is subjected to tests and its accuracy verified with a set of optical fibers that are manufactured or purchased to strict performance requirements. Quite often, a client (say a power utility) purchases a power cable with an integrated optical fiber with very little detail regarding the quality requirements other than to say whether it is a multimode- or a single-mode fiber with its attendant mechanical dimensions.

It is the expectation of the utility that when they connect the purchased DTS system to the power cable, the accuracy, resolution, and longevity are assured. This is generally not achieved unless the power utility has offered to the DTS supplier a length of the exact same fiber incorporated into the power cable and required the DTS vendor to test it on their unit.

For example, a user purchasing a DTS system may place the unit in different ambient environments, say in a substation control room or be ported to the field where ambient air temperature can range from +40 °C to −10 °C. The humidity of the air can also vary. It is important for the user to know that the DTS system selected will yield manufacturer-stated accuracy, demonstrate

Distributed Fiber Sensing and Dynamic Rating of Power Cables, First Edition.
Sudhakar Cherukupalli and George J. Anders.
© 2020 by The Institute of Electrical and Electronics Engineers, Inc.
Published 2020 by John Wiley & Sons, Inc.

reliability, and consistency in the temperature measurements in all these environments. The overall system accuracy is also reliant on the type of optical fiber used and the optical attenuation of the fiber route. Therefore, the users are encouraged to discuss the details of the DTS testing to ensure that the requirements for accuracy, stability, and linearity are achieved.

8.1 Factory Acceptance Tests

Since there are several components that go into the manufacture of a DTS system, in addition to the routine tests of various elements such as laser, fibers, connectors, and control electronics, it is imperative that a factory acceptance test (FAT) procedure be performed. The following sections briefly describe some of the key FAT protocols as well as some special tests. These tests can be performed in the presence of a client if the utility representative is interested in learning how such systems are fabricated, assembled, and tested. Generally, these FATs are witnessed and the certified reports will then be accepted by the utility's representative.

8.1.1 Factory QA Tests on the Fiber Optic Cable

Generally, when considering retrofit installation of DTS systems, especially when there is a spare duct that can be used, it is quite common to select an all dielectric self-supporting (ADSS) cable for the installation. This construction is selected because of the concerns of electromagnetic induction with a metal armored fiber optic cable.

The user will specify the total length of the fiber optic cable for the routes as well as the number of reels that will be allowed for the cable delivery. Generally, the ADSS cable is required to conform to the requirements of the latest IEEE Standard 1222, "IEEE Standard for All Dielectric Self-supporting Fiber Optic Cable (ADSS) for Use on Overhead Utility Lines." The user will also specify the need for a description of the supplier's clamping hardware to clamp cable loops on the ceiling of manholes.

The user may consider including one or more of the following in their technical specification:

- If the cable is intended for circuits requiring a high degree of security, the designed strength of the cable will be greater than the installation may require.
- The number and type of fibers to be contained in the ADSS cable; for example, it may contain six multimode fibers ($50/125/250\,\mu m$) and six multimode ($62.5/125/250\,\mu m$) in separate tubes.
- Whether a loose buffer or tight buffer construction is desired and the requirements for the anti-buckling element shall be specified.

- What should be the nature of the most appropriate water-blocking compound?
- The construction of the cable core elements.
- The requirements for the cable-protective inner jacket.
- The requirements for the outer jacket material and, if there is a need, for anti-tracking material that can withstand induced electrical field strength of 25 kV.
- Confirming that the cable has a zero fiber strain design with the strength elements consisting of two or more layers of torque-balanced aramid yarn applied helically over the inner jacket.
- The requirements for the type of imprinting on the cable with name of the manufacturer, year of manufacture, "OPTICAL CABLE" or an appropriate trade name, and the Standard Optical Cable Code (SOCC).
- The mechanical requirements for any mechanical vertical drop of 10 m without any degradation of the fibers (such as creeping, etc.) over the life of the cable (which is about 30–40 years).
- The nature of the planned installation and the prescription for maximum pulling tension and sidewall pressure for the proposed fiber optic cable. Splicing of the cable in the duct bank is not preferred.
- The nature of requirements for a cable supplier to include such as the product specification, cable construction information, cable diameter, unit mass, predictive modulus or elasticity, coefficient of expansion, maximum rated cable load (MRCL) and minimum bending radius. The cable supplier is required to submit any available documentation which demonstrates the ability of the outer jacket material to withstand long-term environmental and electrical exposure including an electrostatic induction, particularly when this cable has to be coupled to an underbuilt on a transmission line.
- Drawings and requirements for the supplier to provide suitable mounting hardware (such as ceiling clamps, wall clamps that are light, medium, or heavy duty) for the proposed cable.
- Threaded fasteners supplied should be identified as to specification, including type or grade, as applicable with mill test certificates satisfactorily correlated to the materials or products to which they pertain and legible markings on the material or product made by its producer in accordance with the applicable material or product.

8.1.2   FIMT Cable Tests

Factory tests for FIMT with internal multimode/single-mode fibers should include measurement of attenuation at specified wavelength (MM at 850 and 1300 nm, SM at 1310 and 1550 nm, as per the relevant standards). Traditional optical test equipment used in fiber optic communication do not have laser that operates at different wavelengths; hence, the OTDR testing is often done at the above-mentioned wavelengths. The challenge is the interpretation of the optical attenuation (dB/km) and its suitability with DTS equipment that operates at different wavelengths (for example, 1064 nm and 1550 nm). For SM fibers, conventional OTDR do have the 1550 nm test capability. The purpose of this test (with conventional OTDR test equipment) is really intended to confirm the optical integrity of the fiber along the corridor and not to confirm its ability to produce the required accuracy for temperature measurement. It is important to note that using higher wavelength within the window between 1300 and 1550 nm results in lower attenuation (dB/km).

Generally, industry practice is to undertake a two-step optical integrity check on the fiber system. The first OTDR test (with conventional equipment) is used to confirm the integrity of the installation and no damage has occurred to the fiber optic system. The second OTDR test (with DTS equipment) is to verify the optical budget and assess impact on temperature resolution/accuracy.

Some precautions are necessary when installing accessories for underground transmission cables with embedded FIMT. The fibers are very delicate and extreme care must be taken when bringing them out of cables at jointing locations and routing to splicing boxes in manholes. Inevitably, some fiber breakage can occur, so spare fibers are recommended in the design/manufacturing of power cables with embedded fibers.

Before and after the laying of the cable with FIMT, it is recommended to verify the attenuation of the optical fibers to ensure their integrity.

FIMTs typically have a maximum temperature limit of about 90 °C depending on their construction, which means that forming a water-tight seal between sheath and splice body using typical solder wipe methods is not recommended. Instead, an epoxy and fiber-glass tape seal is used, with supplemental embedded copper wires used to electrically connect the splice body to the sheath layer.

It is important for the end user to confirm with the power cable supplier that the incorporated FIMT is capable of withstanding short circuit currents flowing through the sheath of the power cable resulting in sheath temperature reaching 250 °C and this will not compromise the performance of the FIMT.

Since the FIMT is subjected to the same magnetic field as the metallic sheath, there are concerns about the induced losses and their effect on a long-term performance of such FIMT systems. If left "floating," there is a hazard introduced by floating potentials and a risk to operator handling the system while in service.

8.1.3 Temperature Accuracy Test

In this test, the temperature measurement accuracy of the DTS system is verified in each channel; typically at three reference temperatures. Sensing fibers are placed in three ovens where the temperature of the fibers is maintained at three different values and after a steady state is reached, the oven's temperatures are compared with the measured DTS readings. The fiber length in the

(a)   (b)

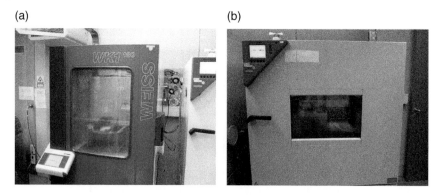

Figure 8.1 (a) Photograph of a DTS system in the oven connected to a fiber loop in a (b) second oven at a specific temperature. (*Source:* courtesy LIOS Technologies.)

oven will be typically 100 m to ensure that enough is available to determine the average temperature and standard deviation, as well as its stability. An example of this test is shown in Figure 8.1.

8.1.4 Temperature Resolution Test

In this test, the temporal noise (standard deviation) of the set temperature reading for each channel is verified using a specific fiber length. The measurement time and ambient temperature of the system are also fixed.

8.1.5 Temperature Reading Stability Test

In this test, the DTS system is connected to a long fiber. For each channel, the temperature at the beginning, middle, and the end of the fibers are verified while the DTS module is placed in an oven where the ambient temperature is cycled between −20 °C and +60 °C. Approximately 100 m of the sensing fibers are placed in the ovens and connected to the DTS system. Ideally, the sensing fiber used in the FAT experiment should be identical in optical characteristics to the one that is installed in the field, but this is not a necessary condition.

8.1.6 Long-Term Temperature Stability Test

For each channel of the DTS system, the overall drift in temperature at the beginning, middle, and the end of the sensing fiber coils is verified over a period of one week while the DTS system is maintained at a specified constant ambient temperature of 20 °C. This will ensure that there is no "drift" in the readings and the measurements appear stable over a certain length of time.

8.1.7 Transient Response Test

From a system operations point of view, it is important that a DTS system responds quite rapidly to the temperature changes. For example, if the load increases or there is a transient change in the loading conditions, the optic fiber and the DTS system should respond rapidly. This becomes critical in road or train tunnels where the initiation of any potential for a fire be immediately detected. Current standards do not define any specific test requirements to prove performance or even provide some idea about the required response time.

It should be recognized that the response of an optical fiber in a heated environment depends on its proximity to the heat source. If the detecting optical fiber is well insulated and forms a part of a cable bundle, further surrounded by a protective jacket and possibly a metal armor, its ability to respond to rapid temperature changes will be contingent on the thermal inertia of the fiber. Nevertheless, the DTS electronics should be capable of detecting such changes rapidly.

In addition to the stated limitations, it is important to note that the response of a DTS is dependent on one or more of the following:

- Number of fibers that are monitored by this unit.
- Desired spatial resolution.
- Number of measurements (averages) to be performed to obtain the required measurement accuracy.

In an attempt to seek a better understanding of possible response times, some utilities have asked a manufacturer to undertake a special test. The requirements were defined to seek a "step-response" of the DTS system. This required a special transient response test to examine how the system reacts to a sudden change in temperature on the fiber. It is well understood that this response will depend on the type of medium surrounding the cable. Therefore, a cable with multiple buffer layers and a FO cable with many fibers may yield a slower response than a cable with a single fiber when subjected to a sudden change in temperature. The test configuration and typical test results are shown in Figure 8.2.

8.1.8 Initial Functional Test and Final Inspection

This test comprises several components:

- Initial function test and final inspection.
- Environmental testing.
- Electromagnetic compliance testing as these systems are placed in a hostile control room environment at a substation.

It is also important to consider what might happen to a DTS system following transportation and final installation. Transportation may result in damage to

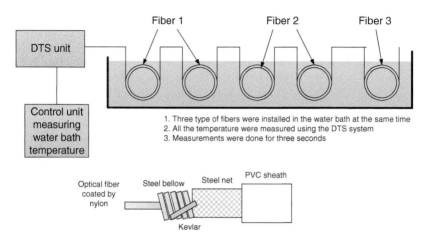

Fiber no. 1:
PVC sheath diameter – 3.0 mm
Steel net diameter: 1.5 mm
Steel bellow diameter: 1.0 mm
Optical fiber diameter: 0.6 mm

Fiber no. 2:
PVC sheath diameter – 2.5 mm
Steel net diameter: 1.5 mm
Steel bellow diameter: 1.0 mm
Optical fiber diameter: 0.6 mm
Jelly injected between steel
net and fiber

Fiber no. 3:
Cable construction in
optical fiber coated by nylon only
Optical fiber diameter – 0.8 mm

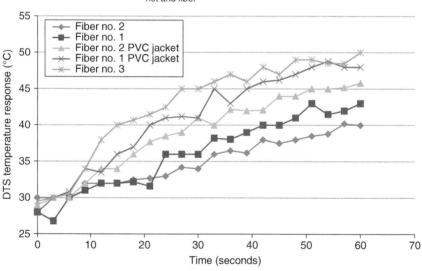

Figure 8.2 Transient response of different optical fibers. (*Source:* courtesy BC Hydro.)

fragile electronics, so mechanical/shock tests can also form one of the investigations in the test protocol. Shock loggers may also be prescribed for safe delivery of these units to the site.

Site acceptance tests (SATs) or final commissioning tests should be done by the DTS supplier in accordance with Installation Instructions and Installation

Inspection and Test Plans provided by the DTS supplier/contractor and accepted by owner/utility's representative(s). They should include field calibration tests.

In an ideal situation and if practical, these site temperature verification tests are done by placing the accessible sections of the fiber ends in water baths with ice maintained at 0 °C and at a higher temperature, say 50 °C, allowing the measurements to stabilize. Usually, for this purpose a length of fiber of approximately 30–50 m should be added at the time of installation at the most convenient locations for this purpose. The DTS measurement should then be compared with independent thermocouple readings of the same water baths. These tests have to be done carefully to ensure that the external ambient temperature does not affect the overall measurements. These measurements should ideally be repeated along the entire cable route.

8.2 DTS Site Acceptance Tests (SAT)

SATs are used to confirm in principle that the equipment and accessories shipped from the factory have reached the site without damage, and once installed, are performing to meet the client's specifications. There are presently no international standards that prescribe what the SATs should be, but it is essential that once the system is installed, the following aspects of the system are verified:

- The DTS provides an OTDR trace that can be recorded for future reference. Ideally, it should be conducted from both ends of the fiber.
- If the sensing fibers are exposed to known temperatures, the DTS system is able to register this accurately.
- The user is able to define zones along the corridor that have some temperature features.
- The defined zones report the set magnitudes (max, min, or average) accurately and reliably.
- Any alarms activated such as high temperature, DTS fail, laser fail, fiber break alarm, and so on, are all invoked and verified to work.
- The SCADA values are being extracted from the external remote terminal unit (RTU) registers with the same quantities (analog values and status alarms) accurately and reliably.
- If the sensing fibers are connected to a DTS calibrator, and when activated to the preset temperature, the DTS reports these values accurately.

Generally, the SAT of the completed DTS system is performed by its supplier witnessed by the utility's employee and the submitted commissioning reports prepared by the vendor is accepted by the utility. Ideally, and if practical, these measurements must be conducted from both ends of the fiber.

The following describes the above principles in some detail and how the tests are conducted in the field.

8.2.1 Final Visual Inspection and Verification of Software Functionality

The equipment, harness, connectors, packing, and so on, are visually inspected. Prior to this, the software functionalities are verified to ensure communication with internal devices, external devices, databases, etc., and are in compliance with the manufacturer's specified reliability and accuracy declarations.

The following sections provide some of the key SAT protocols.

8.2.2 Functionality Test on the DTS Unit

Following the installation of the DTS fibers, power supplies, software, LAN cable connections, and computer, the following are to be verified:

- General setup
- Measurable length
- Spatial resolution
- Sampling resolution
- Measurement time
- Temperature accuracy

8.2.3 Verification of the Optical Switch

For the optical switch (multiplexer), the following items are to be verified:

- Appearance
- Number of input channels
- Number of output channels
- Type of optical controller
- Switching operation

8.2.4 System Control Tests

The system control tests are comprised of the following items that require verification:

- Temperature data acquisition
- Temperature display
- Temperature alarms
- System alarms

8.2.5   System Integration Test with Control Center (if Applicable)

The last phase of DTS integration into system operations involves verification that the DTS-measured data are being stored accurately at the DTS computer, reporting data when polled by the local RTU and transferring the data to the utility's system control center at prescribed intervals and stored in their Energy Management System Database (EMS). It is also important to allow access to the DTS system via VPN-type communication during a trial period to confirm that the data is being acquired and the zones are set appropriately. Thus, if changes are required, there will be no need to travel to site to make these adjustments. It would be prudent for such periodic verification of the system accuracy, reliability of data transferred, and accuracy of the measurements should be planned until a prescribed "burn-in" period is achieved.

8.3   Typical Example of DTS Site Acceptance Tests

The following section describes a sample configuration where a 12 fiber-count ADSS cable was placed between two stations at a utility. Six of the fibers were 62.5 μm MM and the remaining six were 50 μm MM type. The SM fibers were used as communication channels to transfer data back and forth between the two pumping plants along a self-contained fluid-filled cable corridor providing status updates of fluid pressure levels, oil tank levels, and miscellaneous alarms. The 50 μm MM fibers were used to measure duct bank temperatures with a DTS system. The fiber ends were terminated in pigtails with E-2000/APC connectors. A DTS calibrator was placed at one end of this optical fiber system. The original orange pigtails with E-2000/APC connectors were plugged into a DTS calibrator unit to allow reference temperature measurements.

Extra loops of optical fibers for calibration at the sending and receiving ends were prepared to place them alternately in a bath of ice (with a temperature maintained at close to 0 °C) and in a bath of water heated to about 50 °C.

As shown in Figure 8.3, initially, the "calibration loop" close to the DTS unit was placed in ice and hot water and then repeated with the "calibration loop" at the remote end of the route, in ice and hot water sequentially, to adjust the DTS system and thereby confirm that the recorded temperature was accurate, linear, and reliable. Following the measurements and setting of the zones to identify specific features, the DTS calibrator at the remote end was activated to confirm that the DTS system recorded the set points along the corridor. The baths of ice and hot water for calibration are shown in Figure 8.3.

Table 8.1 shows the traces for Channels 1 and 5, where the "hot spots" were isolated into four zones.

Figure 8.3 Water baths for placement of the calibration loops of optical fibers. The stirrer/heater units are held in the bath to make the measurements uniform and a hand-held thermocouple probe was used to separately measure the bath temperature. (*Source:* courtesy BC Hydro.)

Table 8.1 ID of the zones along a cable corridor identifying specific thermal features.

ID	DTS channel ID	Zone name	Metric distances along fiber (m)	Zone ID
1	Channel 2LXX	Zone 1	1170–1240	Between LB and LC
2		Zone 2	4860–5000	Between LK and LM
3		Zone 3	5400–5600	Between LM and LN
4		Zone 4	6640–6950	Between LP and LQ
5	Channel 2LXX	Zone 1	1170–1240	Between LB and LC
6		Zone 2	4860–5000	Between LK and LM
7		Zone 3	5400–5600	Between LM and LN
8		Zone 4	6640–6950	Between LP and LQ

The temperatures were recorded (as shown in Figure 8.4) and the zone temperatures were confirmed to be stored both in the MODBUS registers of the DTS software as well accurately reported to the SCADA monitoring RTU. The DTS unit, computer, and calibrator unit were all powered

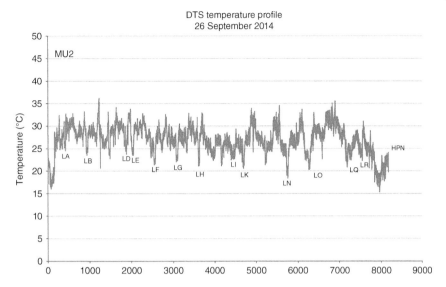

Figure 8.4 DTS temperature profile from site test.

down and powered up to confirm that all alarms were being appropriately displayed and reported via the SCADA. The tests were repeated to confirm that the same data and status were being reported at the local RTU, as well as at the system control center. The final temperature traces showed that most of the hot spots along the corridor reached 30 °C and at two locations (approximately 1200 and 6890 m along the fiber) the temperature reached approximately 35 °C.

Following verification of the DTS system performance, the computer that was being polled by the SCADA RTU was confirmed to receive all the analog and digital values as above and confirmed that all the available activated alarms were initiated and cleared locally at the RTU in the station. Once these were verified and checked, the same was confirmed with the operators at the system control center, verifying that all data were being received during the final end-to-end test. It was decided very early that the RTU will only send an alarm termed "DTS fail" regardless of the nature of the problem, to avoid inundation of information at the control center. Therefore, many of the other DTS-specific alarms were grouped together to annunciate a single DTS fail alarm. If in the future actual loadings approach current ratings, specific zone alarms could be "de-grouped" to aid operators with understanding the alarm causes and appropriate action to be taken. Figure 8.5 shows a schematic drawing of the DTS and fiber system that was installed to monitor the submarine cables at a utility's substation.

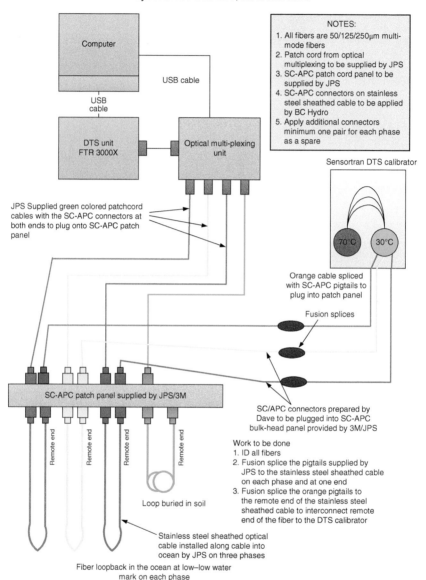

Figure 8.5 Schematic of the DTS fiber layout at all the three substations at a utility.

8.4   Site QA Tests on the Optical Cable System

To ensure a quality product and problem-free installation, the user will normally specify that the ADSS cable shall pass the design tests outlined in IEEE-1222. It will be prudent to ensure that copies of design test results made within the last five years of the following tests be available for comparison. They may include one or more of the following:

Type tests
- Water blocking
- Seepage of filling/flooding compound test
- Electrical tests for Class B
- Test for rated strength of completed ADSS cable using the clamping hardware to be supplied
- Cable twist test
- Cable cyclic flexing test
- Crush test
- Impact test
- Creep test
- Stress/strain test
- Cable cutoff wavelength test
- Temperature cycle test
- Cable aging test
- UV resistance test

The ADSS cable shall pass the following routine tests as outlined in IEEE P1222:
- Jacket thickness
- Cable O.D.
- Optical acceptance tests

Fiber tests
- The optical fibers shall pass the following design tests as outlined in IEEE P1222:
- Attenuation variation with wavelength
- Attenuation at the water peak
- Attenuation with bending
- Temperature cycling
- Attenuation coefficient
- Attenuation uniformity
- Chromatic dispersion
- Mode field diameter
- Concentricity error

- Cladding diameter
- Cladding noncircularity
- Coating diameter
- Fiber tensile proof test

It would be prudent to specify that the cable supplier shall test the reel of cable with an OTDR and record the results. A copy of the results shall be attached to the reel of cable.

Section 8.5 describes the FO sensing cable setup, DTS, and site tests undertaken at the terminal station to commission a Brillouin-based DTS system.

8.5 Site Acceptance Testing of Brillouin-Based DTS Systems

When using Brillouin-based DTS system for temperature measurement, additional site tests are required. These tests can vary depending on a BOTDR- or a BOTDA-based system. The following is an example of a Brillouin DTS response. BOTDA shows different sections along the cable route showing different Brillouin frequencies (a total six in Figure 8.6). This would suggest that there are three different fibers and further they may be from six different manufacturers which will explain three significant and three slightly different frequencies.

Apart from minor variation due to the dispersion in the fiber manufacturing process, we can clearly see three different Brillouin frequencies.

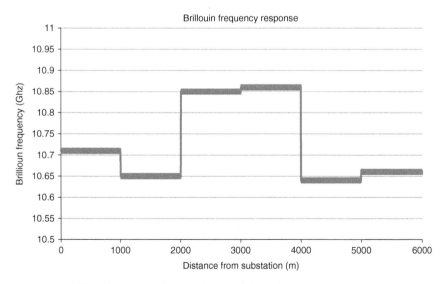

Figure 8.6 Brillouin frequency changes along cable length.

The different Brillouin frequency response has to be smoothed out. During the installation of the Brillouin DTS, a routine operation involves applying a proper calibration coefficient; this will result in the expected continuous temperature profile. If the number of Brillouin frequencies is higher, the response time of the DTS to perform a temperature measurement will be lower.

In order to allow permanent temperature monitoring with a DTS, the characteristics of the optical fiber pair are measured during the FAT test. The Rayleigh response is measured and compared with maximum allowed attenuation. The optical fiber is not tested with the wavelength used by the Brillouin DTS system.

If the required response time of the DTS is short (e.g. five minutes), it is recommended to use the optical fibers with the same Brillouin frequency and to verify the response during the FAT of the cable system.

For integrated FOC in subsea cable system or a separate subsea FOC cable, the Brillouin frequency response has to be verified before layup and loadout to ensure the fibers are useable for Brillouin DTS measurements.

Limits should be set on the strain in the FOC if the Brillouin DTS system is to be used. These should be verified during commissioning.

Two-way OTDR measurements should be used unless there are good reasons not to do so. Optical power meters are necessary where the length of the fiber system is at the OTDR measurement limits.

8.6 Testing Standards That Pertain to FO Cables

IEC 60794-1-2: 2017 applies to optical fiber cables for use with telecommunications equipment and devices employing similar techniques, and to cables having a combination of both optical fibers and electrical conductors. The prime objective of this document is to provide the end user with an overview about the content of different parts of the IEC 60794-1 series numbered -2X. Table 8.2 shows the different parts of this standard.

Table 8.2 Document overview.

Test methods	IEC reference
General guidance	IEC 60794-1-2
Methods E – mechanical	IEC 60794-1-21
Methods F – environmental	IEC 60794-1-22
Methods G – cable elements	IEC 60794-1-23
Methods H – electrical	IEC 60794-1-24

Note: Several numbers in the test method numbering sequence are missing. The reasons for these omissions are historical. To avoid confusion, the existing numbering sequence has been retained.

These documents define test procedures to be used in establishing uniform requirements for the geometrical, transmission, material, mechanical, aging (environmental exposure) and climatic properties of optical fiber cables, and electrical requirements where appropriate. Throughout the documents, the wording "optical cable" can also include optical fiber units, microduct fiber units, and so on. The secondary objective of this document is to provide the end user with useful guidance when testing optical fiber cables. The fourth edition cancels and replaces the third edition published in 2013 and IEC 60794-1-20 published in 2014. This edition constitutes a technical revision. It includes the following significant technical changes with respect to the previous edition and to IEC 60794-1-20:

- The multiple cross-reference tables have been deleted and replaced with a higher level one related to the generic standards; all pertinent text from IEC 60794-1-20 has been included.
- Standard optical test wavelengths have been introduced.
- The document has been streamlined by cross-referencing IEC 60794-1-1.
- The "No change in attenuation" definitions contained in IEC 60794-1-20 have been transferred to IEC 60794-1-1.
- The title has been modified to reflect the contents of the new edition.

This publication is to be read in conjunction with IEC 60794-1-1:2015.

9

How Can Temperature Data Be Used to Forecast Circuit Ratings?

9.1 Introduction

As the demand for electric power is constantly increasing, there is a need to consider options to accommodate the power transfer capacity of the transmission system (transmission lines, cable, and power equipment) to the new requirements with optimal economic and environmental impact.

9.2 Ampacity Limits

The transfer capacity of a cable line is limited to the value of the thermal capacity of the conductor, defined as the static line rating, which is dependent on the soil temperature, soil thermal conductivity, and joule heating in the conductor, sheath, armor, metallic pipes, and the dielectric. Static rating value is determined using parameters defined by the predicted worst-case soil and weather conditions. The static rating is, therefore, a conservative fixed value that would allow operators to overload the lines for short-term emergency operations as defined in the System Operating Orders.

  Since generally, none of the parameters defining the static cable rating are monitored, the power transfer capacity is limited artificially. One possibility to increase the ampacity is to set seasonally dependent static line ratings during which thermal/environmental condition are constant. For example, soil ambient temperature and its thermal resistivity can be changed to fixed values in summer versus winter.

  Transmission cable system involves significant capital investments. Therefore, it is very important to be able to size the cable and design its operating conditions to ensure that a circuit will provide the necessary transmission capacity under the steady state and emergency conditions without jeopardizing its performance while in operation over its entire service life.

Distributed Fiber Sensing and Dynamic Rating of Power Cables, First Edition.
Sudhakar Cherukupalli and George J. Anders.
© 2020 by The Institute of Electrical and Electronics Engineers, Inc.
Published 2020 by John Wiley & Sons, Inc.

Following installation, many a time corridor encroachments occur that can result in potential loss of transmission capacity. If the utility has some posted ratings and these are violated for some reasons, remedial measures have to be taken to restore the current limits to avoid severe penalties.

A transmission cable system is generally surrounded by soil that results in large thermal time constants, which means that the conductor temperature changes can take time even though there are sudden changes in system loading conditions. If the cable system is operating at its capacity, with some intelligent monitoring of the conductor/cable temperature and environmental parameters, quite often additional transmission capacity may be realized. This can reach orders of about 5–10% before the utility needs to consider uprating the cable system, which can prove to be quite expensive.

There are many different ratings that are adopted by utilities some of which are described in Sections 9.2.1–9.2.3.

9.2.1 Steady-State Summer and Winter Ratings

This is the maximum current the cable system can carry in winter and summer continuously over extended periods of time without exceeding the allowable conductor or cable surface temperature. It is important to note that the soil temperatures in winter and summer seasons can vary by as much as 10–15 °C for some utilities and is heavily dependent on the geographical conditions. These ratings also depend on the expected daily load variations.

9.2.2 Overload Ratings

Overload ratings are sometimes requested by system operators to help them determine the ability of a certain cable circuit to carry higher than rated current for a stated period under a contingency condition. For example, if the utility has lost a particular circuit in their system because of a fault, there is a need to ensure service reliability by increasing the loading on another cable circuit without causing it to overheat. These generally are for short duration, say from 2 to 240 hours (~10 days), and would be a typical amount of time it would take to restore a cable circuit following a failure. Transient ratings are synonymous with the overload ratings.

9.2.3 Dynamic Ratings

A step forward is the idea of using dynamic thermal rating for transmission cable operation, based on a real-time monitoring of the current, cable surface and soil temperatures, and using specified or calculated soil thermal conductivity, to arrive at the real-time circuit ratings.

Dynamic thermal ratings implementation would potentially increase the energy transfer capacity by 5–25% above the conventional static ampacity with minimal investments. Dynamic ratings could provide the following benefits:

- Reduce capital expenditures.
- Increase the transmission line rating and reduce the need for new transmission line.
- Increase efficiency – by reducing transmission line congestion constraints.
- Lower customer rates – lower capital expenditures could result in lower customer rates.
- Increase reliability – dynamic thermal rating provides additional information about the available transmission capacity.
- Prevent line rating violation in case of an $N - 1$ contingency situation.

9.3 Calculation of Cable Ratings – A Review

This section offers a brief review of the rating calculations. A detailed discussion of this topic can be found in Anders (1997, 2005). The current-carrying capability of a cable system will depend on several parameters. The most important of these are:

1) the number of cables and the different cable types in the installation under study,
2) the cable construction and materials used for the different cable types,
3) the medium in which the cables are installed,
4) cable locations with respect to each other and with respect to the earth surface, and
5) bonding arrangement.

For some cable constructions, the operating voltage may also be of significant importance. All of the above issues are taken into account; some of them explicitly, the others implicitly, in the rating equations. The lumped parameter network representation of the cable system discussed in Anders (1997) is used for the development of the steady-state and transient rating equations. These equations are developed for a single cable either with one core or with multiple cores. However, they can be applied to multi-cable installations, for both equally and unequally loaded cables, by suitably selecting the value of the external thermal resistance as shown in section 9.6 of Anders (1997).

The development of cable rating equations is quite different for the steady-state and transient conditions. Therefore, we will discuss these issues in Sections 9.3.1 and 9.3.2.

9.3.1 Steady-State Conditions

Steady-state rating computations involve solving the equation for the ladder network shown in Figure 9.1. W_c, W_s, and W_a are conductor, sheath, and armor losses, respectively. The quantity λ_1 is called the sheath loss factor and is equal to the ratio of the total losses in the metallic sheath to the total conductor losses. Similarly, λ_2 is called the armor loss factor and is equal to the ratio of the total losses in the metallic armor to the total conductor losses. Incidentally, it is convenient to express all heat flows caused by the joule losses in the cable in terms of the loss per meter of the conductor. W_d represents dielectric losses the evaluation of which is discussed in IEC 60287-1-1 (2014).

The ambient temperature is the temperature of the surrounding medium under normal conditions at the location where the cables are installed or are to be installed, including any local sources of heat, but not the increase of temperature in the immediate neighborhood of the cable due to the heat arising therefrom. T_1, T_2, T_3, and T_4 are the thermal resistances where T_1 is the thermal resistance per unit length between one conductor and the sheath, T_2 is the thermal resistance per unit length of the bedding between sheath and armor, T_3 is the thermal resistance per unit length of the external serving of the cable, and T_4 is the thermal resistance per unit length between the cable surface and the surrounding medium.

The unknown quantity is either the conductor current or its operating temperature θ_c. In the first case, the maximum operating conductor temperature is given, and in the second case, the conductor current is specified. Since losses occur at several positions in the cable system (for this lumped parameter

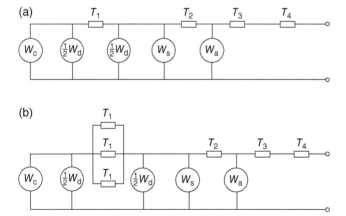

Figure 9.1 The ladder diagram for steady-state rating computations. (a) Single-core cable, (b) three-core cable.

network), the heat flow in the thermal circuit shown in Figure 9.1 will increase in steps. Thus, the total joule loss W_1 in a cable can be expressed as:

$$W_1 = W_c + W_s + W_a = W_c\left(1 + \lambda_1 + \lambda_2\right)$$
(9.1)

Referring now to the diagram in Figure 9.1 and remembering the analogy between the electrical and thermal circuits, we can write the following expression for $\Delta\theta$, the conductor temperature rise above the ambient temperature:

$$\Delta\theta = \left(W_c + \frac{1}{2}W_d\right)T_1 + \left[W_c\left(1 + \lambda_1\right) + W_d\right]nT_2 + \left[W_c\left(1 + \lambda_1 + \lambda_2\right) + W_d\right]n\left(T_3 + T_4\right)$$
(9.2)

where W_c, λ_1, and λ_2 are defined above, and n is the number of load-carrying conductors in the cable (conductors of equal size and carrying the same load).

The permissible current rating is obtained from Eq. (9.2). Remembering that $W_c = I^2R$, we have:

$$I = \left\{\frac{\Delta\theta - W_d\left[0.5T_1 + n\left(T_2 + T_3 + T_4\right)\right]}{RT_1 + nR\left(1 + \lambda_1\right)T_2 + nR\left(1 + \lambda_1 + \lambda_2\right)\left(T_3 + T_4\right)}\right\}^{0.5}$$
(9.3)

where R is the AC resistance per unit length of the conductor at maximum operating temperature.

9.3.2 Transient Conditions

In general, computation of transient ratings requires an iterative procedure. The main iteration loop involves the adjustment of cable loadings. At each step of iteration, the loading of each cable is selected and the temperature of the desired component is calculated. The basic principles regarding transient calculations are laid out in IEC Standard 60853 (1989) and explained in Anders (1997).

Based on the above discussion, the procedure to evaluate temperatures is the main computational block in transient rating calculations. This block requires a fairly complex programming procedure to take into account self and mutual heating, and to make suitable adjustments in the loss calculations to reflect changes in the conductor resistance with temperature. Transient rating of power cables requires the solution of the equations for the network shown in Figure 9.2. The unknown quantity in this case is the variation of the conductor temperature rise with time $\theta(t)$,[1]

---

1 Unless otherwise stated, in this chapter we will follow the notation in (IEC 1989) and we will use the symbol θ to denote temperature rise and not Δθ as in chapter 4 of Anders (1997).

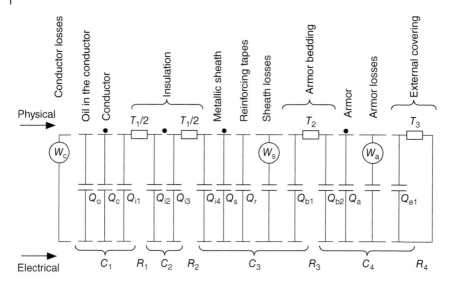

Figure 9.2 Linear thermal network for a general cable with electrical analogy.

Unlike in the steady-state analysis, this temperature is not a simple function of the conductor current $I(t)$. Therefore, the process for determining the maximum value of $I(t)$ so that the maximum operating conductor temperature is not exceeded requires an iterative procedure. An exception is the simple case of identical cables carrying equal current located in a uniform medium. Approximations have been proposed for this case and explicit rating equations developed.

9.3.2.1 Response to a Step Function

9.3.2.1.1 Preliminaries

Whether we consider the simple cable systems mentioned above or a more general case of several cable circuits in a backfill or duct bank, the starting point of the analysis is the solution of the equations for the network shown in Figure 9.2.

Our aim is to develop a procedure to evaluate temperature changes with time for the various cable components. As observed by Neher (1964), the transient temperature rise under variable loading may be obtained by dividing the loading curve at the conductor into a sufficient number of time intervals, during any one of which the loading may be assumed to be constant. Therefore, the response of a cable to a step change in current loading will be considered first. This response depends on the combination of thermal capacitances and resistances formed by the constituent parts of the cable itself and its surroundings.

The relative importance of the various parts depends on the duration of the transient being considered. For example, for a cable laid directly in the ground, the thermal capacitances of the cable, and the way in which they are taken into account, are important for short-duration transients, but can be neglected when the response for long times is required. The contribution of the surrounding soil is, on the other hand, negligible for short times, but has to be taken into account for long transients. This follows from the fact that the time constant of the cable itself is much shorter than the time constant of the surrounding soil.

The temperature-rise of a cable component (e.g. conductor, sheath, jacket) can be represented by the sum of two components: the temperature rise inside and outside the cable. The method of combining these two components makes allowance for the heat which accumulates in the first part of the thermal circuit and which results in a corresponding reduction in the heat entering the second part during the transient.

The reduction factor, known as the attainment factor $\alpha(t)$, of the first part of the thermal circuit is computed as a ratio of the temperature rise across the first part at time t during the transient to the temperature rise across the same part in the steady state. Then, the temperature transient of the second part of the thermal circuit is composed of its response to a step function of heat input multiplied by a reduction coefficient (variable in time) equal to the attainment factor of the first part. Evaluation of these temperatures is discussed below.

9.3.2.1.2 Temperature Rise in the Internal Parts of the Cable

The internal parts of the cable encompass the complete cable including its outermost serving or anticorrosion protection. If the cable is located in a duct or pipe, the duct and pipe (including pipe protective covering) are also included. For cables in air, the cable extends as far as the free air. Analysis of linear networks, such as the one shown in Figure 9.2, involves the determination of the expression for the response function caused by the application of a forcing function. In our case, the forcing function is the conductor heat loss and the response sought is the temperature rise above the cable surface at node i. This is accomplished by utilizing a mathematical quantity called the transfer function of the network. It turns out that this transfer function is the Fourier transform of the unit-impulse response of the network. The Laplace transform of the network's transfer function is given by a ratio:

$$H(s) = \frac{P(s)}{Q(s)} \tag{9.4}$$

$P(s)$ and $Q(s)$ are polynomials; their forms depending on the number of loops in the network. Node i can be the conductor or any other layer of the cable. In terms of time, the response of this network is expressed as:

$$\theta_i(t) = W_c \sum_{j=1}^{n} T_{ij} \left(1 - e^{P_j t} \right) \tag{9.5}$$

where

$\theta_i(t)$ = temperature rise at node i at time t, °C,
W_c = conductor losses including skin and proximity effects, W/m,
T_{ij} = coefficient, K·m/W,
P_j = time constant, s,
t = time from the beginning of the step, s,
n = number of loops in the network,
i = node index,
j = index from 1 to n.

The coefficients T_{ij} and the time constants P_j are obtained from the poles and zeros of the equivalent network transfer function given by Eq. (9.4). Poles and zeros of the function $H(s)$ are obtained by solving equations $Q(s) = 0$ and $P(s) = 0$, respectively.

From the circuit theory, the coefficients T_{ij} are given by

$$T_{ij} = -\frac{a_{(n-i)i}}{b_n} \frac{\prod\limits_{k=1}^{n-i}(Z_{ki} - P_j)}{P_j \prod\limits_{\substack{k=1 \\ k \neq j}}^{n}(P_k - P_j)}, \tag{9.6}$$

where

$a_{(n-i)i}$ = coefficient of the numerator equation of the transfer function,
b_n = first coefficient of the denominator equation of the transfer function,
Z_{ki} = zeros of the transfer function,
P_j = poles of the transfer function.

An algorithm for the computation of the coefficients of the transfer function equation is given in appendix B of Anders (1997).

9.3.2.1.3 Second Part of the Thermal Circuit – Influence of the Soil

The transient temperature rise $\theta_e(t)$ of the outer surface of the cable can be evaluated exactly in the case when the cable is represented by a line source located in a homogeneous, infinite medium with uniform initial temperature. Under these assumptions, the heat conduction equation can be written as:

$$\frac{\partial^2 \theta}{\partial r^2} + \frac{1}{r}\frac{\partial \theta}{\partial r} + \rho_s W_t = \frac{1}{\delta}\frac{\partial \theta}{\partial t}, \tag{9.7}$$

where

ρ_s = soil thermal resistivity, K·m/W,

$\delta = 1/\rho c$ = soil thermal diffusivity (see section 5.4 of Anders 1997), m²/s.

Integrating this equation, we obtain

$$\theta_e(t) = W_t \frac{\rho}{4\pi} \left[-\mathrm{Ei}\left(-\frac{r^2}{4\delta t} \right) \right],$$

(9.8)

where $-\mathrm{Ei}(-x) = \int_x^\infty \frac{e^{-v}}{v} dv$ is called the exponential integral. Its value can be developed in the series

$$-\mathrm{Ei}(-x) = -0.577 - \ln x + x - \frac{x^2}{2 \cdot 2!} + \frac{x^3}{3 \cdot 3!} \cdots.$$

(9.9)

When $x < 0.1$,

$$-\mathrm{Ei}(-x) = -0.577 - \ln x + x$$

(9.10)

to within 1% accuracy. For large x,

$$\mathrm{Ei}(-x) = -\frac{e^{-x}}{x} \left(1 - \frac{1}{x} + \frac{2!}{x^2} - \frac{3!}{x^3} + \cdots \right)$$

(9.11)

The mathematical solution obtained so far is valid under the assumption that the cable is treated as a line source with the internal thermal resistivities equal to that of the surrounding infinite soil. This result is valid for very short times and very deep cable locations only. However, for practical applications, we have to use another hypothesis, namely, the hypothesis of Kennelly, which assumes that the earth surface must be an isotherm. Under this hypothesis, the temperature rise at any point M in the soil is, at any time, the sum of the temperature rise caused by the heat source W_t and by its fictitious image placed symmetrically with the earth surface as the axis of symmetry and emitting heat $-W_t$. With this hypothesis:

$$\theta_e(t) = W_t \frac{\rho_s}{4\pi} \left[-\mathrm{Ei}\left(-\frac{D_e^{*2}}{16\delta t} \right) + \mathrm{Ei}\left(-\frac{L^{*2}}{\delta t} \right) \right],$$

(9.12)

where

D_e^* = external surface diameter of cable, m,

L^* = axial depth of burial of the cable, m.

Under the steady-state conditions, $t \to \infty$ and x approaches zero. In this case, Eq. (9.12) becomes:

$$\theta_e(\infty) = W_t \frac{\rho_s}{2\pi} \ln \frac{4L^*}{D_e^*} \qquad (9.13)$$

As mentioned above, the total temperature rise of a given layer above ambient is a sum of the two parts, where the external part is scaled by the attainment factor $\alpha(t)$ (Anders 1997).

$$\theta(t) = \theta_i(t) + \alpha(t) \cdot \theta_e(t) \qquad (9.14)$$

9.3.2.1.4 Second Part of the Thermal Circuit – Cables in Air

For cables in air it is unnecessary to calculate a separate response for the cable environment. The complete transient $\theta(t)$ is obtained from Eq. (9.2) but the external thermal resistance T_4, computed as described in chapter 9 of Anders (1997) is included in the cable network.

9.4 Application of a DTS for Rating Calculations[2]

9.4.1 Introduction

In the majority of cable installations, maximum permissible conductor temperature is the parameter that limits current-carrying capacity. The limiting temperature is determined by the properties of the insulation material in direct contact with the conductor.

DTS systems provide, in real time, better understanding of how the conductor temperature responds to load variations. The current trend is to measure the temperature at a layer as close as possible to the cable conductor. To this end, the steel tube with fiber optic sensor inside is included between concentric or armor wires or set in direct contact with the jacket, duct, or pipe, depending on the type of cable and installation. In the case of existing installation where the cables are not equipped with the fiber optic sensor, it is a common practice to install the fiber in spare ducts close to the cable.

During the steady-state operation, one can estimate the conductor temperature using the load values, the construction of the cable, and the DTS readings for a given cable layer. The closer the fiber to the cable conductor, the more accurate the estimation is. It is worth exploring the full capacity of the DTS systems once they are installed. The first step would be to install a DTS system to obtain the conductor temperature. However, in practice, the important question is how much more load can the cable carry under steady state, transient, and emergency operations?

---

2 This section is based on the paper by Farahani et al. (2015).

Parameters given in Eq. (9.3) are either constant or change with time. The installation conditions and thermal resistances of the internal layers of the cable are the examples of constant parameters. For underground installations, soil thermal resistivity and ambient temperature are the most important parameters that depend on time. In the case of installations above ground, the ambient temperature, solar radiation, and wind velocity are important time-varying parameters.

The external thermal resistance of an underground cable is directly proportional to the soil thermal resistivity. It can contribute up to 70% of the conductor temperature rise above ambient (Anders 1997). However, its values are largely unknown, even in the steady-state conditions. Measurement at selected locations along the route cannot provide a complete picture of its spatial variations. By the change in season and precipitation, soil thermal resistivity can change in time. Because the space–time variations of this important parameter are normally not known, a conservative approach is used in cable rating analysis and the worst-case scenario is considered. The same conservative approach is used in the case of the soil ambient temperature. Since the conductor electrical resistance is dependent on its temperature, knowing only the conductor current is not sufficient for obtaining the accurate value of the losses and the conductor temperature rise above ambient. If it is the only parameter that is measured, approximate calculations can still be performed with an estimated soil ambient temperature as explained in Anders (2005). The computed conductor temperature can be off by a few degrees in such case.

The brief discussion above should have clarified why one needs to use the DTS systems together with the real-time rating calculations. In a real-time application, the time-varying parameters of the thermal model are updated continuously in time. Using a real-time thermal model one does not need to use conservative approaches and the management of the cable asset is optimized. One can also make emergency and contingency plans more accurately taking into account the real operating conditions of the installation.

There are different approaches to update the time-varying parameters in real time. All these approaches use the load variations and the DTS readings. The criteria to use one approach over another could be the calculation time, required accuracy, type of installation, and, of course, the cost. Ignoring the cost, Section 9.4.2 reviews the existing methods. The following section describes the methods reviewed by Farahani et al. (2015). Several of the methods discussed next were tested and, when available, the experimental setup and the results are discussed as well.

9.4.2 A Review of the Existing Approaches

The objective of this section is to review published approaches that are validated within their defined set of assumptions and limits. It is not meant to compare different methods or comment on their shortcomings or relative advantages but rather to review the state-of-the-art in this field. The focus is on the underground installations.

Anders et al. (2003) assumed the ambient temperature is measured. The load variation and temperature measurements at the cable surface are used to estimate the soil thermal resistivity. The computational algorithm works as follows. Based on known current, cable surface, and ambient temperatures at any point in time t, the soil thermal resistivity is evaluated from the following formula:

$$\rho_{\mathrm{soil}} = \frac{\theta_e - (v-1)\Delta\theta_x - \theta_{\mathrm{amb}}}{nW_c(1+\lambda_1+\lambda_2)+W_d} / vT_4''' \Big|_{\rho_{\mathrm{soil}}=1},\tag{9.15}$$

where $\Delta\theta_x$ is the *critical temperature rise* of the boundary between the moist and dry zones above ambient temperature, v is the ratio of dry and moist soil thermal resistivities (set to 1, if soil dry out is to be ignored), n is the number of conductors, W_d are the dielectric losses, and T_4''' is the external thermal resistance of the cable or pipe. The quantity λ_1 is the sheath loss factor and λ_2 is the armor loss factor. Computation of all these quantities is described in IEC 60287-1-1 (2014). θ_e is the measured cable or pipe surface temperature (it is the temperature between the cable/pipe and soil).

The steady-state external thermal resistance corresponding to this value of ρ_{soil} is given by:

$$T_4''' = T_4''' \Big|_{\rho_{\mathrm{soil}}=1} \cdot \rho_{\mathrm{soil}}\tag{9.16}$$

The temperature rise of the cable/pipe surface above ambient is calculated using an advanced numerical convolution technique. This is done as follows. To compute the temperatures at the present point in time, the value of the soil thermal resistivity is obtained from Eq. (9.16) for a given point of time. This value is used for every integration step i. The following equations describe the integration procedure.

$$\theta_e(t) = \theta_e(\infty) + \sum_{i=1}^{\substack{i\leq m \\ t\leq \mathrm{Time}(i+1)}} A(t,i,i) - \sum_{i=1}^{\substack{i\leq m \\ t\leq \mathrm{Time}(i+1)}} A(t,i+1,i),\tag{9.17}$$

$$A(t,i,j) = W_t(j)\frac{\rho_{\mathrm{soil}}}{4\pi}\left[-\mathrm{Ei}\left(-\frac{D_e^{*2}}{16\delta[t-\mathrm{Time}(i)]}\right) + \mathrm{Ei}\left(-\frac{L^{*2}}{\delta[t-\mathrm{Time}(i)]}\right)\right],\tag{9.18}$$

where $\theta_e(\infty)$ is the pipe surface temperature rise in the steady-state conditions, $W_t(j)$ are the total cable losses at time $\mathrm{Time}(j)$; that is, after the jth integration step, D_e^* is the external surface diameter of pipe, L^* is the axial depth of burial of the pipe, δ is the soil diffusivity, Time(i) is the simulation time (variable integration step), and Ei is an exponential integral function. The number of time steps used in each iteration is denoted by the variable m.

Because the soil thermal resistivity is kept constant during the integration procedure, the computed cable surface temperature at the end of the integration process will be slightly different from the measured one. However, during the next computational step, new soil thermal resistivity will be calculated corresponding to the changing conditions, hence the error will be very small (always below 1 °C). As an alternative, one could recompute ρ_{soil} at every integration step, resulting in a perfect match of the computed and measured cable surface temperatures. Experiments conducted by the authors have shown that this small gain of accuracy does not justify the additional computational effort required.

The computational algorithm takes full advantage of the shape of the measured current, temperatures, and model nonlinearity. This means that the state estimation is not performed at each measurement point (the measurements can be recorded every minute), but the Gear and Adams-Multan predictor–corrector integration algorithms (the time step and iteration order are automatically changed depending on signal variation and required accuracy) are used to obtain desired precision.

Li et al. (2006) introduced the following functional:

$$F = \sqrt{\frac{1}{\tau}\sum_{j=1}^{N}\left[\theta_{est}\left(j\Delta t, \rho_{soil}, \theta_{amb}\right) - \theta_{measured}\left(j\Delta t\right)\right]^2 \Delta t}, \tag{9.19}$$

where τ is the time interval over which the load and temperature readings are recorded for every time step Δt and N is the number of divisions Δt in τ. θ_{est} is the estimated temperature for the point in contact with fiber sensor at time $j\Delta t$. It is a function of ρ_{soil} and θ_{amb}.

For the right choice of the soil ambient and thermal resistivity, the estimated temperature θ_{est} approaches to the measured value $\theta_{measured}$ for each $j\Delta t$ point in time. Therefore, the problem of finding the unknown parameters ρ_{soil} and θ_{amb} is solved by minimizing the functional F with respect to these two time-varying parameters. The finite element algorithm is used to calculate the temperature of the given layer in contact with the fiber.

Brakelmann et al. (2007) extend the equivalent electrical ladder network of the thermal circuit for heat transfer up to the node that defines the ambient. The external soil is divided into sections with their own thermal capacitances C_i and thermal resistances T_i. It is stated that seven sections are sufficient for accurate estimation of conductor temperature variation. Adding more sections increases the calculation time without significant change in accuracy. In this approach, the unknown time-varying parameters are the thermal resistances and capacitances for each section. Once the equivalent electrical network is formed, any appropriate software to analyze electrical circuits can be used. To obtain unknown parameters C_i and T_i of the ladder network, an evolutionary genetic algorithm is used. Although the exact form of the functional is not given, the objective is to obtain the unknown parameters that minimize the difference between the estimated layer temperature and measured values for each instance of time. To this end, a commercial program SPICEOPT, which is based on SPICE, is used. To avoid nonphysical values for the unknown parameters, a search interval is defined.

In a similar approach, Sakata and Iwamoto (1996) use genetic algorithms to estimate lumped elements of the TC sections that represent the external environment. It is reasoned that for short emergency ratings up to six hours, a two-loop network can represent accurately the behavior of the external soil. Therefore, the genetic algorithm is used to update four unknowns in real time. The inputs are the load variation, measured temperature at a given layer using the fiber optic sensors, and the measured ambient temperature. It is stated that conventional minimization to estimate unknown parameters cannot be used. This is because different sets of parameter values yield close minima. Therefore, a genetic approach is used. Other possible methods like extended Kalman filter or simulated annealing are also recommended.

In a semi-analytical-experimental approach, Olsen et al. (2013) have used analytical expressions relating soil thermal resistivity and its specific heat to the soil's moisture content. Although the soil thermal resistivity as a function of moisture content for different soil types show similar behavior, the true dependence for the given soil type at the installation location should be obtained using local measurements. Once the dependence of the soil thermal resistivity and specific heat on moisture content is known, the minimization approach to update the thermal model reduces to that of estimating the moisture content as the only independent parameter. The ambient temperature is measured or is calculated using other analytical or numerical methods. Like the approach used by Brakelmann and Sakata, the external soil is divided into thermal TC sections. However, more thermal sections are used. The internal part of cable is divided into 6 sections and the external part is divided into 100 sections. The other difference is that instead of lumped element parameters T_i and C_i, a single independent parameter, namely the moisture content is updated in real time. To avoid unrealistic estimations of moisture content, a limit on the rate of change of this parameter is imposed.

It should be clear now that transient calculations are the fundamental component of any real-time temperature rating (RTTR) method. The way the effect of the external environment is modelled largely determines the time-varying parameters that should be updated. Existing RTTR approaches differ on the choice of time-varying parameters and the way they are updated. Millar and Lehtonen (2006) and Diaz-Aguiló et al. (2014) provided a review of the transient calculation algorithms.

All methods discussed above have one characteristic in common; namely, by adjusting one or more time-varying parameters they all try to minimize the difference between the measured and computed value of the cable layer temperature. The approach described in Farahani et al. (2015) uses the same principle, but unlike previously described methods, it considers the situations in which the ambient soil temperature is not measured. The approach is similar to the one described in Li et al. (2006) but instead of using finite element method (FEM) to obtain cable layers' temperatures, the latter approach focuses on the application of the IEC analytical method (IEC 60853 1989). As explained in Section 9.3, in the IEC standard (IEC 60853 1989), the temperature rise of any cable layer is divided into two parts, namely:

- The internal temperature rise is the temperature rise of the layer up to the cable surface. This is obtained by solving the ladder network extended up to the surface of cable.
- The external part is the temperature rise of the cable's surface above ambient. An analytic method based on Kennelly's approach and resulting exponential integrals are used to obtain the external response.

The effect of soil thermal capacitance is seen in the exponential integrals by the value assigned to the soil diffusivity. Diffusivity of the soil is inversely proportional to the specific heat. One might decide to consider soil diffusivity as the third unknown parameter to be updated. However, this increases the calculation time in a real-time application without adding much accuracy. One should note that when the thermal diffusivity is not known, the IEC standard (IEC 60853 1989) recommends that the value equal to 0.5×10^{-6} m^2/s should be used. Alternatively, the definition used after Eq. (9.7) relating soil thermal resistivity and its diffusivity can be applied.

All existing RTTR solutions use at least load variation and DTS readings at a given cable layer as the input. There are two steps in obtaining solutions to the steady state or emergency questions defined by the user. In the first step, the unknown parameters that define the thermal model of installation in real time are updated. In the second step, the conductor temperature and the steady state or emergency ratings are calculated. Since in emergency rating calculations one is looking into the future based on present real-time conditions, we need to assume that during the duration of the emergency, the thermal parameters will not change significantly. Of course, one can monitor the response of the cable installation to the emergency load as time passes by to update any changed parameter and take necessary decisions.

In the method introduced by Anders et al. (2003) and Farahani et al. (2015), at each instant of time t, the load variation and DTS readings for the past time interval r are used to update the unknown soil thermal resistivity $\rho_{soil}(t)$ and ambient temperature $\theta_{amb}(t)$ at the present instant t. Depending on the time interval Δt between consecutive load or DTS readings, there are $N+1$ stored measurements for either load or temperature at a given cable layer, $\tau = N \cdot \Delta t$. The load and DTS readings are synchronized. The length of the time interval τ depends on the time constant of the installation. The default is to use 24-hour window. This can be increased for deep installations with larger time constants or when more accuracy is needed. IEC standards for both the steady state and transient calculations are then used to minimize the functional F in Eq. (9.19) for the best solution set ρ_{soil}, θ_{amb}. Once the unknown parameters are obtained, the steady-state rating is used to obtain maximum allowed ampacity based on the real-time thermal model of the installation. This is one of the key objectives of the RTTR calculations. One does not need to use conservative approaches based on the worst-case scenarios and, therefore, the usage of cable installation is optimized.

Transient calculations are used to obtain conductor temperature at time t using the updated thermal model. As time and the 24-hour window moves forward in steps of Δt, the last data point from the input data buffer is omitted and a new data are added. The values for the unknown parameters for the past time t are used as the initial guess in the iterative approach updating the new thermal parameters at time $t + \Delta t$. The next objective of the RTTR calculations is to look into the future for the emergency or contingency plans. Once the unknown parameters and operating point for the present time are known, this is a straightforward application of transient calculations, assuming the load variation is known.

There is a lower limit on the time interval Δt. The limit depends on the length of the cable route and the technology used to analyze the data to provide the temperature profile all along this route. All calculations needed to update the unknown parameters, do the steady-state rating, and calculate conductor temperature or emergency/contingency ratings should not exceed time interval Δt. This is because one needs to deal with the new set of data, load variation, and DTS readings at the end of this time interval. Another practical barrier is that we might want to use one RTTR package to cover many different installations or different thermal sections of the same cable route as the hot spot location may change over time. In this case, all calculations related to all thermal sections should be performed before the next set of measured data for a new point in time for all sections are available.

9.4.3 Updating the Unknown Parameters

The above discussion underlines the need for a fast and efficient method to update the unknown parameters. This part of the analysis consumes most of the calculation time. In an iterative approach, selecting good starting point reduces the calculation time significantly, as the search interval is limited to a smaller domain close to the real solution. The best approach is to use measured soil thermal resistivity at the location of interest. The initial guess for the soil ambient temperature can be obtained by a variety of means. It is recommended to extend at the joints or terminations the same fiber sensors that are used to measure the cable layer temperature, far from the cables, to measure ambient. In practice, it may not be possible to extend the fiber up to 10 m in the normal direction to the cables to make sure that we are measuring the real ambient temperature. In this case, a shorter distance can be selected and the calculations should take into account the fact that some error in ambient measurement is possible.

It is possible to define the search margins around soil thermal resistivity and ambient temperature depending on how much accuracy is required and how much time is available to do all calculations. Although the general principles and approach are the same, depending on the installation type and the available data, the input parameters can be adjusted. For example, for tunnel installations we could use the readings of the fire detection alarms to initialize the value of the ambient temperature. Another possible approach is to use the Kusuda and Achenbach's (1965) formula as explained below.

Assuming a homogeneous earth and that the air temperature at the ground level is periodic, the problem of the heat conduction underground can be solved analytically to obtain the earth ambient temperature as a function of the depth and the soil diffusivity. Kusuda and Achenbach (1965) derived the following relationship:

$$\theta(y, t_h) = A + \sum_{n=1}^{\infty} \alpha_n e^{-y\sqrt{\frac{n\pi}{\delta t_{year}}}} \cos\left(\frac{2n\pi t_h}{t_{year}} - P_n\right) \tag{9.20}$$

in which the phase delay P_n depends on the depth y:

$$P_n = \beta_n + y\sqrt{\frac{n\pi}{t_{year}}}, \tag{9.21}$$

where

t_h = time in hours, h,
y = positive depths relative to the air ground interface, m,
t_{year} = 8766 hours.

α_n and β_n are related to the magnitude and phase of different harmonics of the earth temperature at the interface. The origin of the time can, for example, be the first day of January.

In practice, higher harmonics can be ignored and Eq. (9.21) is simplified to:

$$\theta(y, t_h) = A + \alpha_1 e^{-y\sqrt{\frac{\pi}{\delta t_{year}}}} \cos\left(\frac{2\pi t_h}{t_{year}} - \beta_1 - y\sqrt{\frac{\pi}{\delta t_{year}}}\right) \tag{9.22}$$

Equation (9.22) can be used to calculate ambient soil temperature at any depth. Application of this equation would show that the temperature at a depth greater than about 6 m is equal to the average air ambient temperature A at this location.

Within a good approximation, the amplitude α_1 of the temperature variation at the ground level is close to the average yearly air amplitude, which can be found from the data from the closest meteorological station. Any error is further reduced since the temperature amplitude decreases exponentially with depth. The phase lag β_1 at the air–ground interface is assumed to be close to the phase lag for the air temperature cycle at the given location. The inaccuracy involved would be small in the range of a few degrees. We should remember that this is an approach to obtain a good initial guess of the soil ambient temperature at a given depth.

9.5 Prediction of Cable Ratings

Once the soil thermal resistivity and ambient temperature have been established, the next task is to apply the dynamic rating models to predict short- and long-term cable ratings. This is usually done using the IEC methods described earlier. If the required ratings are longer than about four hours, the question arises what load shapes should be used to model future cable loadings. The most common practice is to use the same shape as was recorded during the last 24-hour period. However, an advantage can be taken from the load forecasting methodology employed by the utility. A brief discussion of this subject is given in Section 9.5.1.

9.5.1 Load Forecasting Methodology

Load forecasting is one of the central functions in power systems operations. The motivation for accurate forecasts lies in the nature of electricity as a commodity and trading article; electricity cannot be stored, which means that for an electric utility, the estimate of the future demand is necessary in managing the production and purchasing in an economically reasonable way. Load forecasting methods can be divided into very short-, short-, mid-, and long-term models. In very short-term forecasting, the prediction time can be as short as a few minutes, while in long-term forecasting it is from a few years up to several decades. This work concentrates on short-term forecasting, where the prediction time varies between a few hours and about one week. Short-term load forecasting (STLF) has been lately a very commonly addressed problem in power systems literature. One reason is that recent scientific innovations have brought in new approaches to solve the problem.

The development in computer technology has broadened possibilities for these and other methods working in a real-time environment. Even if many forecasting procedures have been tested and proven successful, none has achieved a strong stature as a generally applied method. A reason is that the circumstances and requirements of a particular situation have a significant influence on the selection of an appropriate model. The results presented in the literature are usually not directly comparable to each other. A majority of the recently reported approaches are based on neural network techniques. Many researchers have presented good results. The attraction of the methods lies in the assumption that neural networks are able to learn properties of the load, which would otherwise require careful analysis to discover. A vital component of the adoption of dynamic thermal rating is the accurate load forecasting and its implications. This requires continued monitoring of the transmission lines over time (this pertains to both overhead and underground installations), specifically conductor behavior, temperature, and clearance. Rating forecasts are based on statistical analysis using historical data with an assigned forecasting confidence level. Relative change in weather forecast must be also considered, comparing current data to historic weather information. The detailed understanding of the associated variables can mitigate the risk in maintaining safe line operation. Figure 9.3 shows an example of the amount of an overhead bare conductor line capacity versus % of time with this overload capability using different forecasting tools.

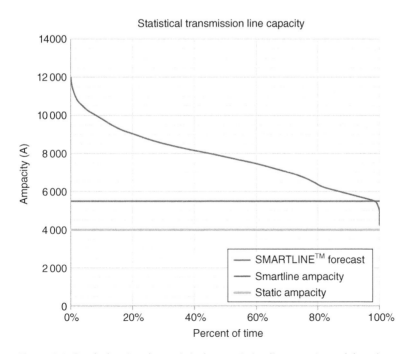

Figure 9.3 Graph showing the statistical transmission line capacity and the advantage of improving the forecast capability.

A good forecasting tool should provide the following:

- accurate 24 hours ahead prediction
- the thermal limits must not be exceeded
- provide frequent automatic update of the line capacity

It is important to note that the load forecasting is based on statistical analyses of historical data to determine circuit rating or overload capability while the dynamic line rating (DLR) makes use of the prevailing load information and ambient weather conditions to establish the dynamic rating for the circuit. It is important for the operator to trust the prediction and operate in the most conservative way using static security limits when prediction is compromised.

Several years back BC Hydro utilized an artificial neural network (ANN) based system to forecast dynamic ratings. In principle, if one has a population of data that show calculated ratings for prevailing weather conditions, it is possible to develop dynamic ratings using ANN models by using forecasting techniques. The ANN model obviously will have to be "trained" on a set of data to allow accurate predictions under a different set of ambient conditions. While the technique appears to be a good mathematical tool, the challenge remains for the cases when the real data is completely outside of the training set of environmental parameters.

9.6   Software Applications and Tools[3]

RTTR gives valuable information about the overload capability of the cable system rather than just the cable conductor temperature. A system operator is not concerned about the temperature and is also unsure as to where the temperature is being measured along the cable corridor and how it relates the power transfer capacity. They are also oblivious to the impact surrounding soil may have on the cable temperature and thus unsure about the relationship between the maximum load current and the DTS readings. From an operator's standpoint, some of the considerations with the deployment of such a system may be:

- What is my transfer limit on the stated circuit at this instance?
- What is the overload capacity for this line under a contingency condition?
- How much time do I have before the cable temperature limit is exceeded at the present overload condition?

The real-time rating systems can be used to answer these questions. The calculations require application of complex computation algorithms and, therefore,

---

3 The description of the features of the RTTR programs is based on the material obtained from the vendors. Whereas the CYME and EPRI programs have been tested by the authors, the remaining ones were not.

advanced software tools. Generally, each solution provider offers software applications and tools that are available to process field data and calculate the optimal dynamic or quasi-dynamic thermal ratings of the transmission circuits. Figure 9.4 shows a schematic of typical RTTR software.

The architecture and the manner in which the results of the real-time data make their way on to the utilities, energy management system (EMS), vary. Some utilities prefer that all the computations be made locally at the DTS computer and only the ratings be posted into the EMS system. On the other hand, some utilities prefer to get the DTS data into the EMS system, and the control center will use the tools to undertake ratings calculations and integrate the "data intelligence" into the EMS system to allow the system operators to use the "intelligence" to optimally operate the network. In summary, the need to implement a dynamic rating system also referred to as real-time thermal rating

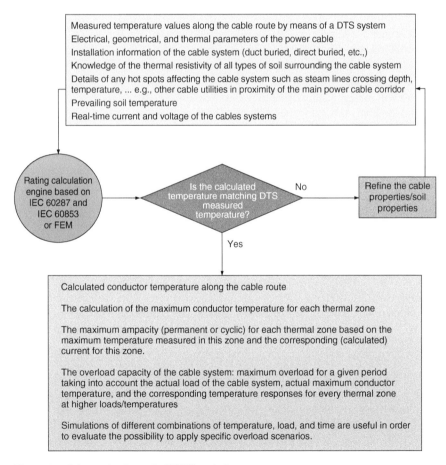

Figure 9.4 Schematic of a typical RTTR analysis.

(RTTR) will be dictated by the operators' needs from a system flexibility, resilience, and reliability point of view. Each case and circuit has to be carefully validated with proper business justification. Some of the key elements to be considered when trying to decide on which philosophy to adopt may be best served by developing the framework after answering the following questions:

- Is the cable asset near its operating limit today and does it have additional margin to allow temporary overloading of the circuit?
- Is there a risk of having reached the operating limit of the asset?
- Is there appropriate DTS data to allow its integration, load data, and dynamic analyses data into the EMS system?
- Who will validate the results from the DTS and RTTR systems and confirm that accuracy over the long term is maintained?
- Where should the RTTR software reside and who will install, validate, and maintain the program? Should the RTTR software be placed close to the DTS system, should be at the control center of the utility or system operator?
- Is it important to integrate the DTS computer system and the RTTR software?
- Should only intelligently processed data be sent to the EMS at the control center?
- Who will develop the system architecture and ensure security compliance, communication protocols?
- Who will maintain upgrades to the real-time ratings software by vendors?
- Who decided what kind of communication protocols should be adopted between the RTTR, DTS, and the EMS systems?

Sections 9.6.1–9.6.4 provide a description of several RTTR tools that are commercially available to utility engineers who are operating their assets near their operating limits.

9.6.1 CYME Real-Time Thermal Rating System

CYME Int. is one of the leading software providers engaged in cable rating business. Their real-time thermal rating (RTTR) software is based on the IEC standard to update the thermal model of cable installations in real time as discussed in Section 9.4. Their model has been tested on different types of installations to demonstrate that one can use DTS readings, load variation, and the IEC standards to calculate conductor temperature and optimize the usage of cable asset in the steady state and transient situations.

CYME's RTTR takes advantage of a DTS system to measure the temperature profile along a cable corridor that is then used to obtain the temperature values at the "thermally governing section." These values are used to calculate the circuit ratings (amperes or MVA) in real time using this DTS temperature, load currents, and soil thermal resistivity data. If the soil parameters are unknown, the functional (9.19) is minimized over the preceding 24 hours period to estimate their values. The way the accuracy of the calculations has been verified is discussed in Sections 9.6.1.1.1–9.6.1.1.2.

9.6.1.1 Verification of the Model

9.6.1.1.1 *Test Setup*

During the prequalification test for one of the cable manufacturers performed on $2500\,mm^2$, enameled copper conductor, 420 kV cables, the entire cable route was equipped with 480 m of compact external fiber optic sensor cable to monitor the temperature profile by a Raman-based DTS instrument. The sensor cable was a dielectric single gel-filled tube with $8 \times 50/125\,\mu m$ multimode fibers, with a diameter of 6.5 mm and a weight of 35 kg/km. The prequalification test was done according to the IEC Standard 62067. The test arrangement covered five different installation conditions:

1) Cables installed in open air.
2) Cables directly buried.
3) Cables installed in a steel-plated shed above the ground
4) Cables installed in a non-ventilated PE tube direct buried.
5) Cables installed in an underground concrete tunnel.

Some of the installations are shown in Figure 9.5.

The test loop was heated by conductor current to a given temperature. The heating was applied for at least 15.5 hours. The conductor temperature was maintained within the stated temperature limits for at least two hours at the end of each heating period. This period has been followed by at least 32.5 hours of natural cooling. The total duration of 1 cycle was 48 hours.

Hundred cycles of heating and cooling have been carried out with a conductor temperature 0–5 °C above the maximum in normal operation (90 °C). Eighty cycles of heating and cooling have been carried out with a conductor temperature of 105 °C (emergency temperature). A voltage of $1.7U_0$ has been applied to the parallel test loop during the whole test period (8760 hours).

To assess the ambient temperatures, which were used for the RTTR calculations, loops of the sensor cable were deployed at various positions as shown in Figure 9.6.

The test setup involved also identical reference cables to the cables installed on the main loop of the test in an almost voltage-free setup, permitting installation of thermocouples on the conductor as recommended in the standard. This is schematically shown in Figure 9.7.

These reference cables were installed close to the main loop to be in the same thermal conditions. This made it possible to actually measure the conductor temperature for comparison with the RTTR results. The results for the section where the cable had been installed in a non-ventilated PE tube buried directly underground are discussed in Section 9.6.1.1.2.

9.6.1.1.2 *Results of the RTTR Calculations for the Test Installation*

DTS technologies permit evaluation of the entire temperature profile of the deployed test cable. Figure 9.8 shows the temperature profiles of the cable test

Protective tube   Sand – directly buried

Cable tunnel   Cable tunnel with joints

Figure 9.5 Four installations at which the RTTR system was tested. (*Source:* with permission from CYME Int.)

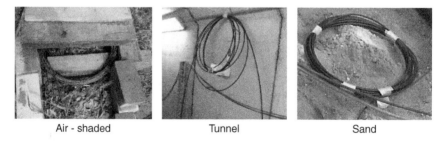

Air - shaded   Tunnel   Sand

Figure 9.6 Illustration of the ambient temperature measurement at various installations. (*Source:* with permission from CYME Int.)

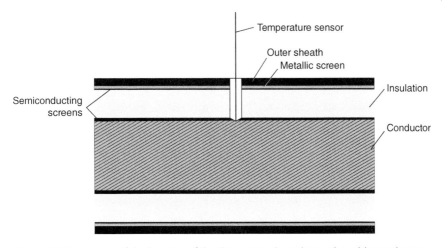

Figure 9.7 Illustration of the location of the thermocouple probe on the cable conductor. (*Source:* with permission from CYME Int.)

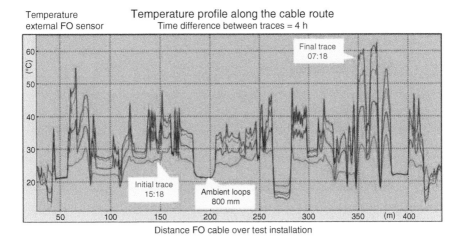

Figure 9.8 Temperature profiles along the cable route. (*Source:* with permission from CYME Int.)

loop at different times from one outside cable termination to the far-end termination, including the various installation conditions.

Figure 9.9 shows ambient temperature during one heating cycle. It follows the seasonality – the ambient temperature inside the unventilated tunnel shows in addition a strong relation to each heating cycle.

From the available constants (cable type and laying condition) and the dynamic parameters (load, DTS readings, soil resistivity, and ambient), the

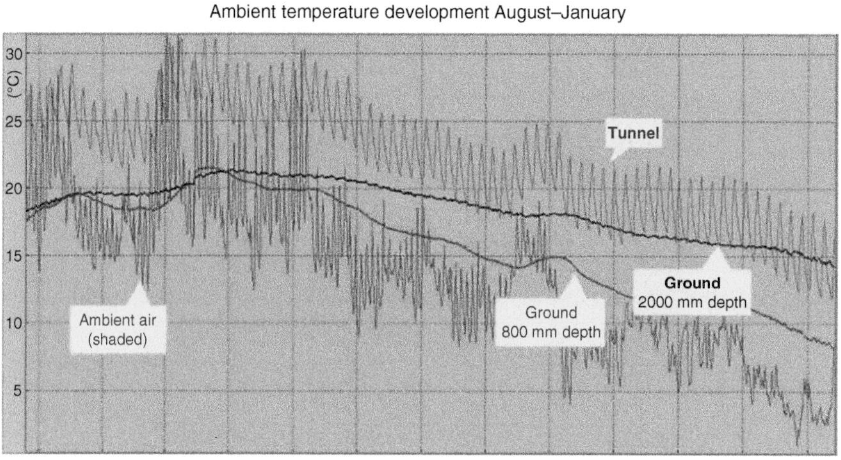

Ambient temperature development August–January

Figure 9.9 Ambient temperature measured during heating cycles. (*Source:* with permission from CYME Int.)

conductor temperature was calculated and is just slightly above the measured value, as shown in Figure 9.10.

9.6.2 EPRI Dynamic Thermal Circuit Rating

EPRI began applying distributed fiber optic temperature sensing (DFOTS) beginning in the mid-1990s to member utilities looking to optimize, and potentially increase, the ratings on existing underground cables or to provide as-built verification of the design ratings on new cable circuits. In these early days, there was only one supplier of DTS equipment and the equipment was fairly expensive. At that time, EPRI purchased a unit for testing at cable-using utilities.

Though there have been several more commercial installations and several DTS vendors, many utilities still face uncertainties and challenges to the successful application of DFOTS technology to cable systems. EPRI responded to this challenge by developing an application guide to help utilities in assessing the selection and application of DFOTS technology.

The goal of this guide was to help its membership develop an improved understanding of these systems and take away some of the uncertainty of using DFOTS with underground power cables. This was done by describing how the equipment works, and the current state-of-the-art of fiber-based temperature measurements, discussing the benefits of different types of installations of DFOTS equipment, highlighting some of the challenges to using the technology and a means to interpret the results. The main purpose and goal were to educate and guide utilities in the evaluation and application of the technology. The authors of the guide began contacting DFOTS equipment suppliers to get

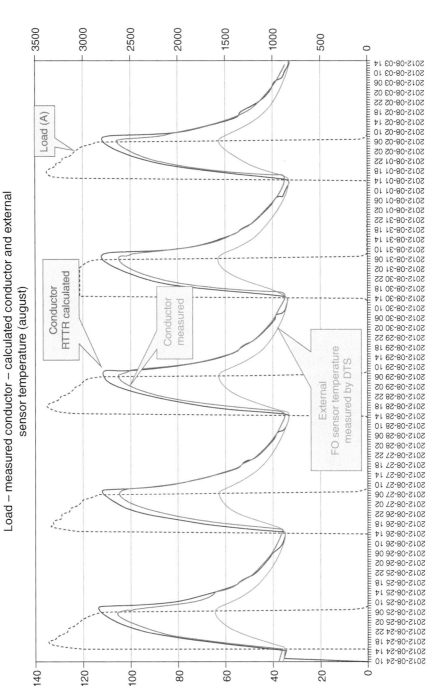

Figure 9.10 Comparison of measured and computed conductor temperature for an underground installation. (*Source:* with permission from CYME Int.)

a general understanding of the state-of-the-science. This information was combined with EPRI's background information on DFOTS systems from the various EPRI projects that have utilized DFOTS methods as a basis for the case studies. A general implementation strategy is provided so that engineers at utilities can make more informed decisions about applying the DFOTS technologies.

Following the success of applying sag-tension monitoring devices and weather stations to rate overhead lines, EPRI developed and field-tested a software tool called dynamic thermal circuit rating (DTCR) that calculates dynamic thermal ratings based on real-time monitoring of weather conditions, line loading, sag, and temperature to rate overhead lines. They also developed module to rate power transformers by using line current, ambient air temperature, and transformer top and bottom oil temperatures. This software received considerable enhancements with the passage of time where the capabilities to rate other power equipment was incorporated. Presently, this EPRI DTCR system is not only capable of dynamically rating overhead lines but can also rate other series-connected equipment such as transformers, power cables, CTs, high-voltage bus, and so on. It is unique in this regard. Just as line models were defined to accept weather-influenced data, transformer models were also developed to use top and bottom oil temperatures, soil temperature along power cables, etc., to help rate all the equipment and provide an operator with the governing rating along a specific transmission corridor. There are many examples of utilities adopting this tool, particularly for overhead lines, and realizing increased asset utilization in the order of 5–15% at relatively low cost. The tool also helps to avoid conductor damage from excessive heating. At BC Hydro, the early versions of the DTCR software was used to rate several power transformers in their network. Enhanced versions of the software were used to develop dynamic ratings for a 525 kV submarine transmission cable by using distributed temperature measurements data of the actual conductor. The conductor temperatures could be measured since the optical fiber was placed inside a hollow conductor as described in Chapter 5.

A disadvantage of this software system is that it is complex. It needs the user to customize the various elements in the network, deploy additional hardware such as thermocouples to measure ambient soil temperatures, top and bottom oil temperatures sensors, undertake in-situ measurements of the soil thermal resistivity, and validate the accuracy of the data before it can be incorporated into the program and validated. One must also have sound knowledge of the module that is being used for analyses and to understand which of the available models to use in the context of the application.

9.6.3   DRS Software by JPS (Sumitomo Corp) in Japan

JPS (now Sumitomo Corporation) has supplied a Raman-based DTS application and also offers dynamic circuit rating (DRS) system software to rate transmission cables. The company has more than 15 years of operating history with such software.

Their DRS model is based on the IEC 60287 and IEC 60853-2 where cylindrical thermal resistance and capacitances are assumed. Engineers at JPS (Igi et al. 2011; Watanabe et al. 2015) demonstrated examples of the field installation trial measurements conducted for two of their clients at the laboratory and in the field on an actual 330 kV land cable system used in a project in Australia and a 138 XLPE cable in an existing pipe of a HPFF cable system for a project in New York.

Their dynamic real-time monitoring can provide conductor temperature alarm output at the predetermined levels, compute allowable maximum overload display for preset time durations based on present conductor temperature in real time, identify the hottest spot recorded by the DTS system, and provide conductor temperature distributions in real time. This system is also capable of predicting the conductor temperature rise, peak load, or time duration to reach maximum allowable temperature. It can also examine archived cable surface and conductor temperature histories up to five years at elected sites along the cable route. It can also provide trend graph displays of stored data for designated duraion in intervals of 24 hours, 1 day, 1 month, 1 year and 5 years.

The authors also conducted some experiments by placing the DTS sensor on the power cable surface in a concrete encased duct bank along with a DTS cable above the duct system to determine the relationship between the DTS temperature near and away from the cable surface.

In the first paper (Igi et al. 2011) they describe how the DRS system records and displays the conductor temperature profile and activates alarms according to the conductor temperatures of 90, 100, and 105 °C. Their DRS system can also provide statistical data for cable asset managers/system operators by showing the cumulative hours of cable conductor temperature range for some installation conditions as shown in Figure 9.11.

The thermal resistivity of the soil is computed by this system from the standard Fourier equation:

$$\rho_s = \frac{\theta_e - \theta_{amb}}{W_{tot}}, \tag{9.23}$$

where

ρ_s = soil equivalent thermal resistance, K·m/W,
W_{tot} = mean value of the total heat dissipated from cable, W/m,
θ_e = mean cable surface temperature, °C,
θ_{amb} = mean ambient soil temperature, °C.

Application of Eq. (9.23) requires the knowledge of the soil ambient temperature. In Igi et al. (2011), the authors described how they buried a length of the

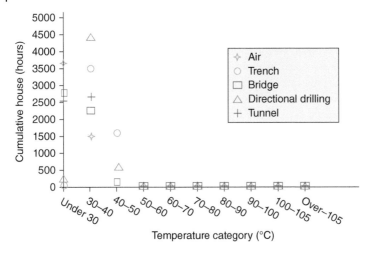

Figure 9.11 Statistical record display of the JPS DTS system. (*Source:* with permission from JPS.)

Figure 9.12 Ambient temperature sensor. (*Source:* courtesy of JPS.)

optical fiber at the same depth and farther away from the current-carrying cable as shown in Figure 9.12 to measure the ambient soil temperatures.

Figure 9.13 shows a typical screen from their DRS software where the user is able to describe which prediction is needed.

9.6.4   RTTR Software by LIOS

A DTS manufacturer in Germany offers a real-time rating software to complement their DTS measurements. It calculates the conductor temperature profiles, identifies critical locations, and triggers (pre-) alarms when the respective thresholds for the conductor temperature are reached or exceeded. The full software package called EN.SURE RTTR offers prediction capabilities. All

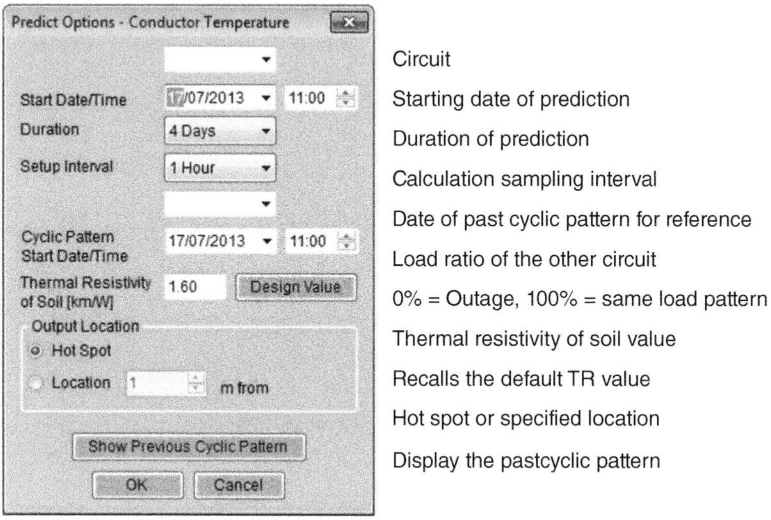

Figure 9.13 Typical prediction scenario requirements.

thermal models used in the rating are validated by FEM calculations to ensure the accuracy for various load scenarios. It is fully integrated with the user-friendly and easy-to-learn database and visualization software. All readings from multiple DTS units and point sensors as well as the rating results are stored together in a modern SQL database. A complement of visualization tools enables multiscreen visualization of temperature profiles and histories, current data, and intuitive, custom visualization of all data on maps, pictures, or drawings of the full power cable installation.

This software is capable of considering different types of cable configurations as shown in Figure 9.14.

A DTS vendor participated in a study at a European laboratory to demonstrate the effectiveness of their DTS and RTTR systems and thereby establishing its accuracy by comparing the calculations with actual measurements. The test involved the monitoring of a cable system during prequalification tests on a 320 kV HVDC cable with 2500 mm^2 Al conductor, XLPE insulation, aluminum sheath, and a HDPE jacket. The cable sections were placed in different installation conditions, namely tunnel, directly buried in soil, open air, PVC duct, and directly buried in sand. A DTS system was used to monitor the power cable outer sheath temperature (continuous/linear profile). They applied their RTTR calculation engine to calculate the conductor temperature (continuous/linear profile) in real time and predict conductor temperature in an emergency. This was then compared with the DTS/RTTR results using a third-party reference temperature sensor equipment. Figure 9.15 shows an example of comparative measurements using a classical and a DTS method for certain loading conditions.

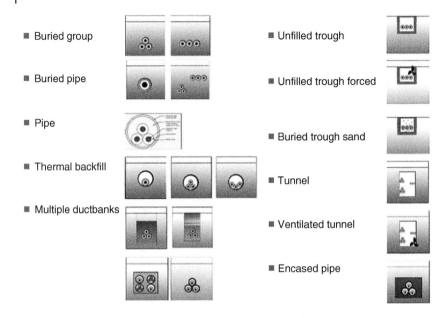

Figure 9.14 Illustration of different cable configurations the Real-time rating software is capable of handling.

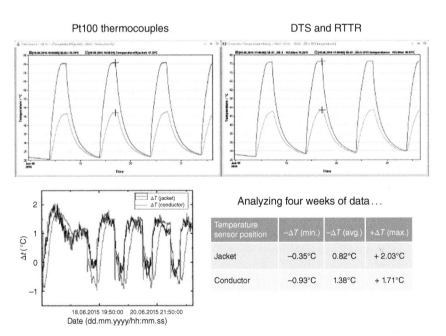

Figure 9.15 Comparative measurements using classical and DTS measurements. (*Source: courtesy of LIOS Technologies.*)

Enhanced visualization and RTTR predictions

- Enhanced visualization …
  - (Rogowski coil) current sensor data
  - DTS fiber and conductor temperature profile chart
  - Fault (hot spot) color representation

- Rating summary …
  - CT (conductor temperature) for real-time steady-state calculation of the conductor temperature profile based on DTS temperature, current data, and other relevant thermal parameters. *PVC duct is the limiting thermal section*. However, hot spot is at 11 m (open air –> tunnel)!
  - RTTR for real-time transient calculation of the conductor temperature profile and emergency predictions based on OTS temperature, current load history (profile), and other relevant thermal parameters.

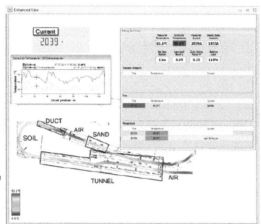

Figure 9.16 Comparative measurements using classical and DTS measurements – computer screen display. (*Source:* courtesy of LIOS Technologies.)

During the measurements, it became obvious that due to various installation conditions the "thermally limiting section" was the installation of the cable in a PVC duct. This served to determine the governing rating for the test circuit. Figure 9.16 illustrates some features of the software.

9.7 Implementing an RTTR System

Implementing an RTTR system requires capital investment (computer, software, and SCADA) in addition to the investment made for the DTS itself. The distance capability of the DTS system will depend on whether the installation being monitored is a land or a submarine cable. Spatial resolution needs imposed on the DTS system for the land cables will generally be more onerous compared to the submarine installations. In case of a submarine cable, the DTS system selected will require longer distance measurement capability. All these parameters would ultimately govern the cost of the DTS system.

The cost impact of the RTTR system will depend on whether or not a single RTTR software is capable of using data from several DTS systems possibly from different vendors. The costs will also be governed by the ability of implementing the future extensions (e.g. extra RTTR servers in Master/Slave configuration).

In order to obtain this information, additional installation requirements could be necessary such as: placement of thermocouple/fiber optic "trees" to measure soil temperature, placement of current transformers (CTs) to track the real-time current of the cable, and placement of additional thermal probes (thermocouple, fiber optic) along the cable corridor especially at hot spots and splicing locations. The nature of communication links between those dispersed sensors and the main DTS/RTTR system will dictate the response time of the system.

The cyber security aspects of the RTTR system installation and distributed sensor data have to be taken into consideration.

The modeling of a particular RTTR system can take several weeks to prepare the software for implementation provided that all the power cable system information is known.

9.7.1 Communications with EMS

The RTTR system could enable communication with the SCADA of the grid operator. In this case, no direct communication to a RTU will be possible; all necessary inputs will be acquired from the EMS.

Typical protocols for communication with SCADA system are:

- IEC-60870-5-104
- IEC-60870-6/TASE2 (ICCP)
- Modbus RTU/ASC/TCP
- DNP3
- IEC 61850

To convert the standard protocol of the server to the SCADA, protocol software or hardware convertors can be used. Software converters are installed directly on the RTTR server while hardware converters are separate electrical components.

In order to evaluate the communication, it is important to specify the frequency of data transfer from the RTTR system to EMS and vice versa taking into account what the demands are for the available bandwidth, whether the throughput from the RTTR to EMS is constant or works in peaks.

Figure 9.17 gives an overview of the main network components.

The contractor will specify the necessity of a continuous throughput between the RTTR and EMS. The software should be made resistant as much as possible for the consequences of lost data or interruptions on the communication link. In case of loss of a communication link, the impact on the reliability and start-up time of the RTTR system should be clear.

As an example, consider the following question: What happens with the RTTR system if the DTS connection is offline? Typically, the RTTR software can continue to perform the calculation without the verification of the DTS measurement. Because the DTS temperature is not available anymore, any change in the cable environment will affect the accuracy of the results. The

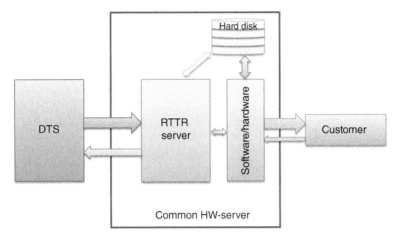

Figure 9.17 Example of network connections.

time required to reach the same level of accuracy as before the interruption will depend on whether the missing DTS temperatures are sent to the RTTR. If the missing data is received, the required accuracy can be reached in a short period of time (hours). If the missing data is not available, the time to reach the desired accuracy will be longer.

9.7.2 Communications with the Grid Operator

The typical protocols used for LAN/WAN networks of grid operators in substations are:

- Protocol: Ethernet and TCP/IP.
- Interface: optical fiber, 10/100 Mbps, SC connectors.

For each device connected to the LAN (DTS, data concentrator, information treatment unit, etc.), it is important to have an estimation of the data volume and the frequency of the data transfer on the LAN.

9.7.3 IT-Security, Data Flow, Authentication, and Vulnerability Management

In order to evaluate data transfer, a description of the dataflow should be given describing which flows are established between each component of the solution (e.g. port, protocol, sender, receiver). A description how the integrity of the data in these flows can be assured helps to estimate the impact on the vulnerability of the IT system.

To use a RTTR system in a control room, different roles are assigned to engineers, IT-specialist, and end users. Every role uses the information from the RTTR system differently. For example:

- Dispatchers control the balance and load of the grid and are mainly interested in normal and emergency load values of the cable systems.
- Engineers are mainly interested in the thermal behavior of the cable system (e.g. hot spots) in order to evaluate the possibility of removing the hot spots and thermal bottlenecks.
- IT-specialists are interested in the dataflow, data security, and communication path between different components.
- Asset managers will make use of RTTR for future investment grid planning taking into account overload scenarios.

A logging should be centrally accessible with at least logging of user access and configuration changes present.

There should be a process in place where any vulnerabilities should be mentioned directly in the system; for example, in case of a problem in data communication. In such a case, for instance, all connected peripheral devices (USB, CD ROM) should be disabled in order to avoid misuse.

9.7.4 Remote Access to the RTTR Equipment

Approved users should have access to the RTTR equipment from a remote location only if they first make a connection to a secured network (via secured VPN tunnel). From there, a connection to an internet page of the equipment at the substation level is possible through a connection via a specified jump server. Special attention should be paid to allow direct access from a remote location for the external users.

9.8 Conclusions

It is evident that utilities are becoming increasingly interested in harnessing DTS and DRS systems to alleviate transmission congestions and optimize the utilization of their assets. The numbers of DTS suppliers are increasing and the versatility and software features are expanding while the costs are decreasing. The challenge today is that while it is easy to adopt the use of the DTS systems on new cable systems, it is not so easy to apply on existing installations since the placement of fibers poses a challenge. There have been examples of fiber being placed in a spare duct in a ducted cable system, but the relationship between the measured DTS fiber and the inferred cable conductor temperatures is not easy to establish.

DRS systems are also being offered by a limited number of vendors but they rely not only on the measured temperature but also on real-time data of load currents, soil temperatures, and soil thermal resistivity to make accurate prediction of the ratings without compromising cable performance.

Several demonstration systems have been applied under controlled test laboratory conditions to validate the predictive performance under various installation and loading scenarios. However, the experience on actual real-time systems and used by system operators is limited.

A good RTTR system is characterized by a complex and accurate computational algorithm and has several important practical features. The system can be applied to all voltage levels, various cable constructions and installations, including: paper (high- and low-pressure fluid filled), extruded, gas, in air, directly buried, submarine, ducts, and cables in tunnels. The minimum input to the calculations is the load current and cable (or pipe) surface or other layer temperatures obtained from the fiber optic measurements. The program should be able to compute not only the steady-state and emergency ratings but also provide information on the time required to attain the specified temperature.

RTTR systems can be used to enhance the current-carrying capacity of power cables as well as to eliminate risks of overheating. They allow utilization of cable systems to their maximum capabilities and are particularly useful when a deferment of the costly capital programs is desirable.

10

Examples of Application of a DTS System in a Utility Environment

This chapter contains several examples of the application of a DTS system in BC Hydro. In early 1990s, the utility undertook an extensive program of monitoring some of the more heavily loaded transmission cables. This, most likely, was the first such attempt in the world and it provided a valuable experience for the future similar endeavors around the globe.

10.1 Sensing Cable Placement in Cable Corridors

BC Hydro is an electrical utility that has an extensive 230 kV underground system in a metropolitan area. Some of these older circuits periodically operate near their thermal limits under first contingency conditions. Majority of the cables are of a self-contained fluid-filled (SCFF) construction and installed in duct banks consisting of four 125 mm I.D. ducts, one of which is meant to be an empty duct for a spare cable. If some cables appear to be heavily loaded, the utility would like to "push" the limits of operation on these underground circuits under summer peak situation. In this transmission network, two underground circuits are particularly heavily loaded in the summer and these transmission cable systems are thermally limiting. In the past, the utility had very little information about the cable conductor, sheath, or duct bank temperatures to ensure the thermal limits are not violated.

In an effort to defer expensive system reinforcement projects, a variety of ampacity uprating methods can be investigated and sequentially applied, in order to incrementally increase the safe loadability of the desired circuits. This phased approach aids the preparation of business cases for each stage of improvement, with early work aimed at filling planning and engineering needs. But if there were a means of providing real-time rating information to control center operators, this could result in more efficient use of the transmission system.

Distributed Fiber Sensing and Dynamic Rating of Power Cables, First Edition.
Sudhakar Cherukupalli and George J. Anders.

Normal ampacity ratings were based on classical calculation methods and generally conservative assumptions had been made about the cable losses and thermal environment, for either "summer" or "winter" conditions. This could have included a reduction of 10 °C in the normal maximum allowable conductor temperature of 85 °C, to allow for possible unknown hotspots along the cable route. This factor of safety conforms to North American high-voltage cable industry standard specification AEIC CS4. In our example, emergency overload ampacities were simply based on normal ampacities calculated at a 20 °C higher conductor temperature, but only for short durations, and did not incorporate transient temperature calculations.

With the prospect of expensive system reinforcement projects to relieve apparently overloaded cables, beginning in about 1994 efforts were made to implement methods to establish cable ampacities more accurately and less conservatively. This included application of the EPRI's dynamic thermal circuit rating (DTCR) system. The DTCR system provides dynamic normal and short-time ratings in real time, based on historical and predicted loadings and actual ambient temperatures. Thermal models were provided for most equipment in a typical transmission line, including overhead lines, cables, transformers, and so on.

The other new development was the introduction of the distributed fiber optic sensing technology. If the utility had means of incorporating an optical fiber in this 2×2 duct bank to allow monitoring the temperature along the corridor, the spare duct temperature could be recorded in real time. With the help of conventional mathematical model, one would then be able to calculate the spare duct temperature at a given load and compare the calculated results with the DTS measurements to determine how good a match one obtains. If the match is good, the exercise can be repeated at other load conditions. Once the mathematical model was found to be consistent and robust, one could use the same to predict the circuit overload capability and then use the information from the DTS system to ensure that temperature limits are not violated.

There are several utilities around the globe, which have commenced harnessing this technology in various sectors either to isolate hot spots along cable corridors, address fluctuating power transfers through underground and subsea cables from large offshore wind parks, the presence of other utilities such as large distribution circuits, steam line intersecting these corridors, or generally deteriorating soil conditions along the cable route.

10.2 Installation of the Fiber Optic Cable

In order to help retrofit into an existing cable corridor, the availability of a spare duct is important. If the route has several manholes along the corridor, it would be prudent to consider interconnecting the spare duct route at each end in an effort to make the retrofit installation efficient. Such a retrofit was undertaken at BC Hydro in 2002. Figures 10.1–10.3 show several stages of installing an optical cable in a spare duct. Figure 10.1 shows the air compressor in the background with the air-line feeding a "super air jetting unit" in the foreground. Figure 10.2 shows a close-up view of the blowing unit with the air-hose connected to the top with a control valve. The orange duct connected on the side of the "air jetting unit" is used to direct the fiber optic cable into this system. Figure 10.3 shows the reel of the all-dielectric self-supporting (ADSS) fiber optic cable being led into the "super air jetting unit."

Prior to blowing the FO cable, the ribbed conduit has to be pulled into the spare duct. Figure 10.4 shows how a ribbed plastic duct is being pulled into a manhole. The number of ribbed ducts introduced into the spare duct is a function of number of cables to be used. Quite often, since the DTS fiber is usually a multimode $50.125/3000\,\mu m$ type, this sensing fiber is blown into one duct. The second telecom cable, which is generally a single-mode fiber construction with a large fiber count (90 or 144 fibers) is blown into the second duct and the two FO cables are kept separate.

Figures 10.5–10.7 show the photographs of the ribbed ducts being pulled into the spare duct along the concrete duct bank. It has been shown that when ribbed ducts are used to pull into existing ducts, they are easier to pull since the contact surface area is reduced lowering frictional resistance.

Figure 10.1 Connection of air jetting equipment and a ribbed duct connecting into a fiber duct in a manhole. (*Source:* courtesy of BC Hydro.)

(a)

(b)

Figure 10.2 (a) Close-up view of the Superjet equipment with the ribbed duct attached to guide the FO cable and the air-line connected to the top with a valve; (b) close-up of the ribbed duct. (*Source:* courtesy of BC Hydro.)

Figure 10.3 Fiber optic cable reel supplying the optical cable to the spare duct. (*Source: courtesy of BC Hydro.*)

Figure 10.4 Photograph showing two ribbed ducts being pulled into a single duct. (*Source: courtesy of BC Hydro.*)

Figure 10.5 Photograph of a FO cable being blown into one duct and preparation for a second one to be introduced into the second duct. (*Source:* courtesy of BC Hydro.)

Figure 10.6 Photograph showing how excess cable is placed between the pulls in a figure-of-eight configurations when long blow lengths are encountered. (*Source:* courtesy of BC Hydro.)

(a)   (b)

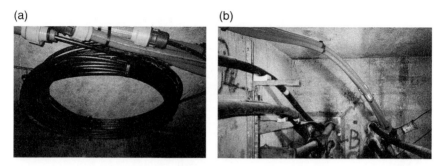

Figure 10.7 (a) Coupling of ribbed ducts inside a manhole to allow continuous installation of the optical fiber and (b) the exit of three subducts out of a single spare duct inside a manhole. (*Source:* courtesy of BC Hydro.)

10.3 Retrofits and a 230 kV SCFF Transmission Application

BC Hydro decided to proceed with a demonstration DTS project on a 230 kV circuit between two stations. This is in the study area shown in Figure 10.8. DTS measurements were made using a 50/125 μm multimode optical fiber that was installed in a spare duct for the communication path from the host DTCR PC at Station 1 to remote terminal units (RTUs). An eight-fiber cable was selected so that both DTS and DTCR communication from several RTUs could be accommodated. A low-cost "plenum" cable was used and it was installed with only one splice.

10.3.1 Early 230 kV Cable Temperature Profiling Results

Figure 10.9 shows a September 1996 DTS profile for the entire route, at a time when the circuits had been operating at about 85% of normal maximum ampacity for several days. Much of the cable route is along a grassy right-of-way. Asphalt road crossings and parking areas are clearly hotter. Manholes are cooler. Air temperatures were about 16 °C at the time of measurement, as shown by the FO cable temperature at an RTU instrumentation box where it was routed above ground near Manhole GI. This agrees with the air temperature shown at Hill Ave. Terminal Station is where the FO cable was routed above ground to an equipment cabinet and fibers spliced to form a continuous temperature sensing loop. Ambient soil temperatures at the same locations are shown to be about 19 °C.

New information was gained on the location of previously unknown hotspots in the first section from Newell Substation (Figure 10.10), at Arcola Street (Figure 10.11) and at Canada Way (Figure 10.12). Some were suspected to be due to the presence of sections of high thermal resistivity soils and

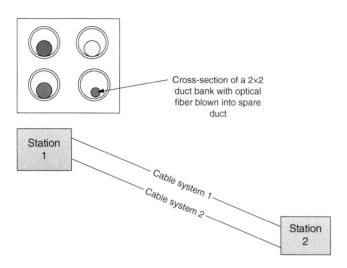

Figure 10.8 230 kV underground transmission study area.

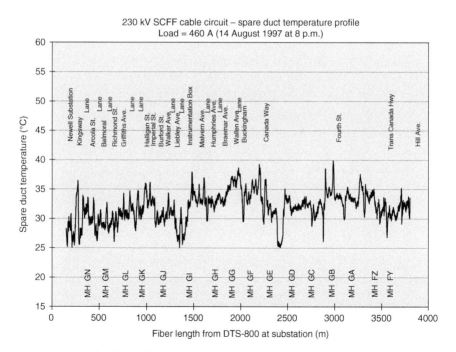

Figure 10.9 230 kV SCFF cable – spare duct summer temperature profile for the entire route.

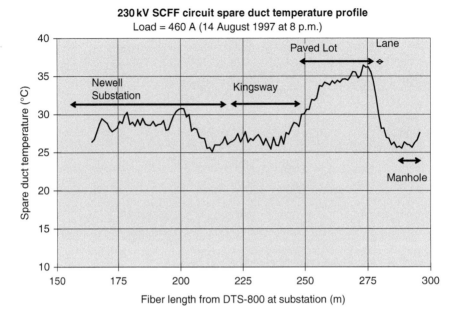

Figure 10.10 230 kV SCFF cable – spare duct summer temperature profile – Kingsway hotspot.

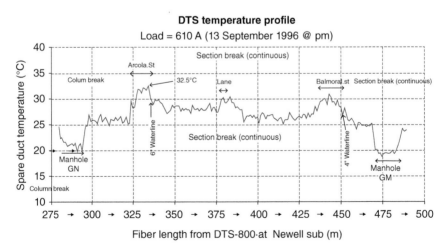

Figure 10.11 230 kV SCFF cable – spare duct summer temperature profile – Arcola-Balmoral hotspot.

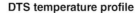

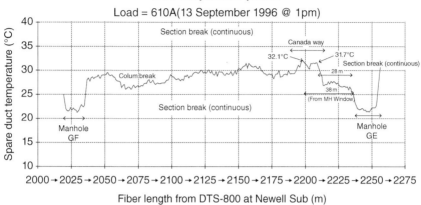

Figure 10.12 230 kV SCFF cable – spare duct summer temperature profile – Canada Way hotspot.

backfills. Others, such as at Canada Way, are caused by the excessive burial depths (about 5 m), which took place long after initial installation. This location is now inaccessible for remedial excavation due to busy traffic.

Water pipe locations are evident at Arcola and Balmoral Streets, corresponding with the city records. Because of cooler manhole temperatures, the fiber optic cable was purposely looped around the inside of each manhole in order to provide convenient reference points on the profile plots. These are clearly visible in the profiles. Having extra FO cable in the manholes will also help to refine future instrumentation.

Cable heating seems usually to be highest in the summer where cables are under asphalt. This is caused by the combined effects of solar heating, lower soil moisture contents directly beneath the road, and lower heat transfer convection coefficients at the asphalt surface.

Figures 10.13–10.15 show January '96/97 temperature profiles for the same sections of the cable route, when loadings had been about 55% of the normal maximum for several days. Cable heating under asphalt surfaces is not as high, but the Canada Way hotspot is still visible because of the deep burial condition.

10.3.2 Location, Mitigation, and Continued Monitoring of the 230 kV Hot Spots

In winter, the dominant hotspot was shown to be at Canada Way. Eventually, the duct/cable voids could be filled there with removable grout in order to improve heat transfer. Local water circulation in the spare ducts in the 2L39 and 2L40 duct banks is also a possibility.

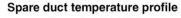

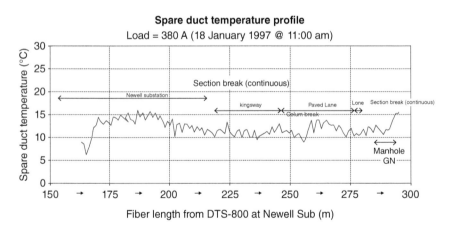

Figure 10.13 230 KV SSCF cable – spare duct winter temperature profile – entire route.

Figure 10.14 230 kV SCFF cable – spare duct winter temperature profile – Kingsway hotspot.

In summer, the hotspots at the first section from the Newell Substation and at Arcola St. were shown to be marginally hotter than the Canada Way hotspot (33.5 °C compared to 32.1 °C). Because it was present throughout the year and considering the closeness of both locations, it was decided to concentrate on the Canada Way hot spot as the limiting location.

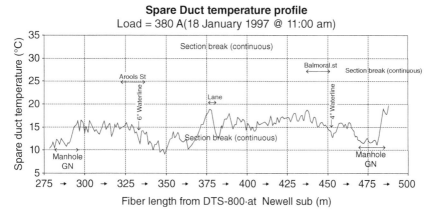

Figure 10.15 230 kV SCFF cable – spare duct winter temperature profile – Arcola-Balmoral hotspot.

As mentioned before, it is most desirable to measure cable temperatures as close to the conductor as possible. This reduces error margins due to assumptions about thermal resistivities and also reduces temperature response time after a change in the load, since there is less thermal impedance (capacitance and resistance) between the heat source and the measurement point. This awareness is important with the installation of optical fiber in the spare duct rather than in the same duct as the power cable. However, the ease of retrofitting the DTS made this relatively insignificant shortcoming acceptable. Subsequent investigations showed that measuring only the spare duct temperature was sufficient to overcome the above concerns. Further investigations have shown that the steady-state cable conductor temperature could be inferred quite accurately, given the steady-state spare duct measured values.

10.4   Example of a DTS Application on 69 kV Cable System

Sometimes, the portability of the DTS systems allows them to be installed to monitor several circuits. The following is an example of such application on a 69 kV cable system. As part of a re-rating effort on three 60 kV underground circuits between two major downtown substations, FO cables were pulled into spare ducts in two separate duct banks in order to determine the thermal "health" of the SCFF cables. Figure 10.16 shows the temperature profile obtained for one of the duct banks containing two circuits.

A previously unknown major hotspot location was found at a steam heat line crossing. As shown in Figure 10.16, the spare duct temperature was over 50 °C.

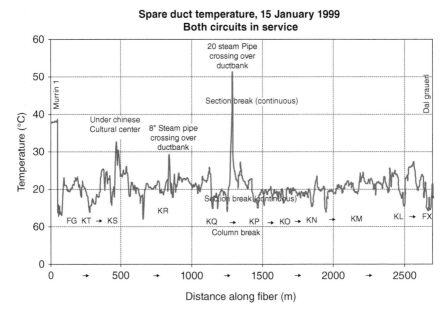

Figure 10.16 69 kV – spare duct temperature profile showing steam heat pipe hotspot.

Information gained by using the DTS helped mitigating the hotspot, determining the extent of insulation aging that had occurred at this location, calculating new circuit current ratings, and formulating plans for local cable replacement. Without the DTS systems, the hotspot may have gone undetected for many more years, possibly leading to premature cable failures and undesirable customer outages.

10.5 Verification Steps

Investigations were carried out to compare the DTS measurements with calculated cable component and spare duct temperatures and thermocouple measurements. The approach was to determine if the DTS records in the spare duct could be used to accurately predict the cable conductor and jacket temperatures.

Calculations were done using off-line analytical approach, based on the IEC 60287 standard, off-line finite element analysis methods, and real-time calculations done using EPRI's DTCR program. The latter system was installed to provide dynamic ratings to a BC Hydro area control center, under a separate project.

10.5.1 Analytical Methods

A custom Mathcad ampacity program was prepared, based on current rating calculation methods described in the IEC Standard 60287, with daily load factor effects modeled as per Neher/McGrath paper. This allowed calculation of the component temperatures for each cable, as well as those for all ducts, given various conductor currents, daily load factors, soil ambient temperatures, soil resistivities, and so on. As part of the DTCR project, a RTU with thermocouple inputs was installed near Malvern Street, where indicated on the temperature profiles in Figures 10.9 and 10.13. The cable installation is identical to that shown in Figure 10.8, except that the top of the duct bank is 1.60 m deep. Thermocouples were used to measure ambient soil, duct bank surface, spare duct, and cable surface temperatures. The RTU was integrated with the DTCR system to give open-loop feedback information on actual cable surface temperatures and ensure accurate calibration of the DTCR thermal model.

Based on the Mathcad analysis, and referring to Figure 10.17, it was found that there was close agreement between the calculated and DTS-measured spare duct temperatures at the Malvern Street location, once the thermal model was adjusted using an inferred soil thermal resistivity of 0.57 K·m/W. However, the measured cable jacket temperatures using two separate thermocouples were found to be about 10 °C smaller than calculated, and even smaller than the measured spare duct temperatures using DTS. This anomaly could only be explained by ground water leaking into the duct bank at the location where it was broken into to attach the thermocouples. Ground water flowing downhill within the cable/duct voids could also artificially lower the cable surface thermocouple readings. There is evidence of the effect of ground water flowing down cable ducts elsewhere along the route, for example, between Manhole GE and GD, where the duct bank is on a 3° slope and wet soil conditions prevail.

We can observe from Figure 10.17 that measured spare duct temperatures using DTS are on average much more reliable indicator of the thermal state of the cable system compared to discrete temperature sensors, such as thermocouples.

Similar methods were used to calculate the conductor temperatures that would result at full loading, at various locations along the route, based on the summer and winter DTS spare duct temperature profiles (see Figure 10.18). This was used as a basis for determining realistic parameters, such as soil thermal resistivity, to be used in calculating the "book" ratings to be used for the System Operating Orders, as well as for the DTCR thermal model. The hotspot location at Canada Way turned out to be the limiting thermal section most of the time. Therefore, it was used as the main control section for the entire circuit. Ultimately, implementation of the fiber optic cable reel supplying the optical cable to the spare duct, the DTS and DTCR led to higher normal and short time current ratings than those previously used, with a high degree of confidence.

General conditions	Malvern	Arcola
Depth to top of ductbank (m)	1.6	1.6
Daily load factor	0.9	0.9
Load (A)	450.0	450.0

Measured temperatures (thermocouples)	Malvern	Arcola
Soil temperature (°C)	13.4	13.4
Top of duct bank (°C)	19.6	N/A
Cable 1 jacket temperature (°C)	21.8	N/A

Measured temperatures–DTS fiber	Malvern	Arcola
Spare duct temperature (°C)	24.0	29.0

Calculated temperatures	Malvern	Arcola
Cable 1 conductor temperature (°C)	40.80	44.30
Cable 1 sheath temperature (°C)	33.40	36.90
Cable 1 jacket (°C)	32.00	35.50
Cable 1 duct mean temperature (°C)	28.70	32.30
Cable 1 duct inside temperature (°C)	25.50	29.20
Spare duct temperature (°C)	25.20	28.90
Inferred coil thermal resistivity (°C-m/W)	0.57	0.75

Figure 10.17 Measured and calculated temperatures at two locations.

10.5.2 Dynamic Thermal Circuit Ratings

In parallel with this project, BC Hydro applied the EPRI DTCR system to the three circuits identified in Figure 10.1, one of which was monitored using a DTS unit. At that time this was the most extensive application of the DTCR to underground cables among all EPRI members.

The preceding tabulation of calculated cable component temperatures shows that the inside temperatures of occupied ducts are marginally greater than the spare duct temperature, at about 60% of full load. At full load, the spare duct temperature is calculated to be only about 3 °C smaller than in the duct carrying the cable. Thus, by comparing the calculated and continuously archived cable duct inside temperature with DTS-measured values, a high degree of confidence was gained in the results of the DTCR dynamic rating calculations, which are used by system operators on a daily basis.

This "manual open-loop feedback system" is an effective way of achieving reliable dynamic ratings, but a higher level of sophistication could be achieved by directly integrating DTS systems with the DTCR. This could be done by outputting time/temperature/distance data from the DTS system to DTCR via its RS-232 port, partitioning into spatial/temperature zones associated with

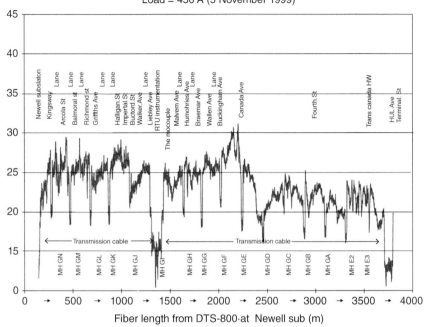

Figure 10.18 230 kV SCFF cable – spare duct temperature profile 5 November 1999.

different thermal environments of the cable circuit, and using these data as a basis for cable conductor temperature tracking and dynamic ratings. Such a system has been implemented for the shore ends of the Gulf of Aqaba 400 kV submarine cables (Balog et al. 1999) using FO cables permanently attached to the cable exterior.

Figures 10.19–10.21 demonstrate the applications and benefits of dynamic ratings compared to static ratings, as implemented with DTCR on the three 230 kV underground circuits at BC Hydro.

10.6 Challenges and Experience with Installing Optical Fibers on Existing and New Transmission Cables in a Utility Environment

Application of distributed temperature sensing methods to optimize the use of power cables can now be considered a mature technology. However, each application on existing lines presents challenges with retrofitting the fiber optic cables. Temperature profiles require careful evaluation to understand the

DTCR - Thermal Ratings Report				
		Circuit	Circuit Elements	
		◄ unit ►	◄ unit ►	
Date: 07/15/99		2L40	Newell Malvern Canada_Way	
Time: 09:40			b1150UG60 b1150UG90 b1150UG180	

		2L40	Newell	Malvern	Canada_Way
			b1150UG60	b1150UG90	b1150UG180
Normal	Amps	714.0	799.9	766.7	714.0
100 Hr.	Amps	912.0	980.7	954.0	912.0
24 Hr.	Amps	972.5	1047.0	1017.7	972.5
8 Hr.	Amps	1064.9	1148.0	1115.3	1064.9
Circuit Load	Amps	344.0	344.0	344.0	344.0
Time To Overload✱	Minutes	N.A.	N.A.	N.A.	N.A.

✱TTO computation based on Normal

Description
2L40 from Barnard Tap via Hill Avenue Terminal

◉ Amps   ○ Hours   Cap Bank MVAR   0.0

○ MVA   ◉ Minutes   Load Power Factor   1.0

[OK]   [Update]   [Details]   [Msgs]   [Help]   ☑ Automatic Update

Figure 10.19 Screen shot of dynamic thermal circuit rating output report for three different locations (duct bank cross-sections) 9 : 40, 15 July 1999.

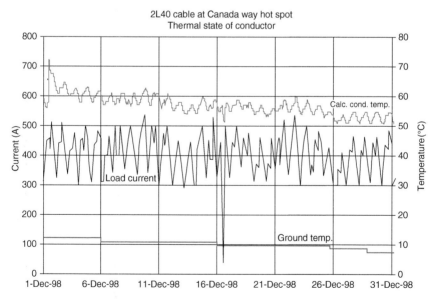

Figure 10.20 230 kV dynamic thermal circuit rating system with calculated thermal state of the conductor, December 1998.

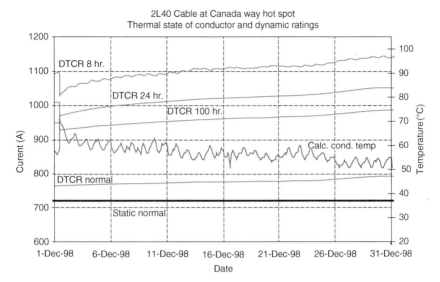

Figure 10.21 230 kV circuit – dynamic thermal circuit rating system with calculated dynamic ratings compared with static rating, December 1998.

three-dimensional thermal processes experienced by the cables, under a variety of loading conditions and external effects, throughout the year.

BC Hydro has achieved very tangible benefits from applying DTS technology to power transmission cables, and expects these to continue as more applications arise.

In summary, the following positive results were achieved, either as a direct part of this project or in conjunction with the associated work.

- Commercial distributed temperature sensing equipment was researched, specified, purchased, lab tested, and successfully field proven.
- Temperature profiles were obtained for one 3.5 km 230 kV and three 2.7 km 69 kV underground transmission lines, for a variety of load and seasonal conditions.
- Temperatures measured with the DTS sensors were compared with traditional discrete measurements and the superiority of the DTS for providing reliable thermal diagnostics was proven.
- Improved confidence in the 230 kV cables' thermal environment resulted in higher normal and short time current ratings, compared to those previously used.
- DTS measurement results were used to accurately and reliably calibrate the thermal models used in EPRI's dynamic thermal circuit rating system, when applied to existing 230 kV cables in ducts.

- Insight was gained into the methods and limitations of integrating DTS measurements with dynamic rating systems using optical fibers in nearby spare ducts.
- Previously unknown severe hot spots were found, where steam heat lines crossed existing 69 kV underground transmission cables.
- Methods were developed to insert optical fibers into the conductor cores of the existing 525 kV submarine cables to facilitate distributed temperature sensing in the thermally limiting shore sections.
- New knowledge was gained about the thermal behavior and performance of the underground transmission lines in a variety of environments.

Any large power utility with an extensive underground transmission system should benefit from application of DTS technology in similar ways.

In summary, this chapter showed how temperature monitoring of existing transmission corridors may be conducted by blowing fiber into spare ducts and isolating hot spots along the route. Two studies, one of a 230 kV system and a second 69 kV system, were presented and how the impact of hot spot on future performance was also discussed.

11

Use of Distributed Sensing for Strain Measurement and Acoustic Monitoring in Power Cables

11.1 Introduction

As discussed in the preceding chapters, fiber optic cables (FOCs) are now quite commonly incorporated into the power cables for temperature measurements. However, there are many more opportunities for harnessing the FOCs. Sections 11.2 to 11.9 describe distributed strain sensing and distributed acoustic sensing applications of the fiber optic sensor for power cable applications.

11.2 Strain Measurement

Recently, distributed strain monitoring (DSM) systems are arriving into the market place. Such systems offer a significant benefit if they are applied on a submarine cable. When the cable leaves the barge or a ship and is dropped into the sea, it is subjected to high tensions. Without proper control, the cable could be damaged; hence, the installation tensions need to be monitored. Real-time monitoring of strains or tensions during installation and operation of the submarine power cables is essential in order to make qualified decisions regarding safety and reliability of the line.

During installation, tensions and curvatures of the cable can be calculated for various sea states if the catenary configuration and the response characteristics of the installation vessel are known. If the layback distance between vessel and touchdown point is too short, the power cable can be damaged due to overbending and local kinking. By increasing the layback distance, the top tension must be monitored closely to ensure that there is no overloading of the cable or equipment. The strain measurement would also help identify the exact touchdown point.

Distributed Fiber Sensing and Dynamic Rating of Power Cables, First Edition.
Sudhakar Cherukupalli and George J. Anders.
© 2020 by The Institute of Electrical and Electronics Engineers, Inc.
Published 2020 by John Wiley & Sons, Inc.

Because of the random nature of the wave loading, the installation planning involves running a large number of dynamic simulations of load cases to ensure that the mechanical integrity of the cable is not compromised. The decision to start or abort an installation campaign is, therefore, based on the predefined limiting sea states for each stage of the operation. If real-time monitoring of the cable strain was available, the knowledge of the actual tensions and curvatures in the cable system may allow for increased confidence when deciding when to start or abort the cable lay. This could have significant cost savings if it meant abandoning the installation or not during an inclement weather. The identification of the origin of a damaging stress is crucial to assign the responsibility to the cable manufacturing or installation (when they are different). Damage to cable during installation is best avoided considering the high repair costs.

11.3   Example of Strain Measurement of a Submarine Power Cable[1]

11.3.1   Introduction

BC Hydro recently installed several three-core power distribution cables with embedded fibers, connecting the mainland grid to Bowen Island, British Columbia, Canada. Including an optical fiber in the distribution cables provided opportunities not only for telecommunication and temperature measurements using a distributed temperature sensing (DTS) system but also proved useful in strain measurement via a DSM system.

While there have been significant studies that delve into the use of such fiber optic systems in laying pipelines, their benefits remain undetermined with regards to power cables. This chapter discusses recent studies completed at BC Hydro's subsidiary test laboratory that aim to understand the sensitivities of such a system, and identify some of the challenges surrounding its implementation on a real submarine cable application might offer. To deduce strain from an optical glass fiber, the physical phenomenon of Brillouin scattering is utilized.

Brillouin scattering is an inelastic scattering of light in the presence of acoustic phonons (lattice vibrations) in the propagation medium. The frequency of acoustic phonons is a material property that varies as a function of temperature and strain. This means a measurement of this frequency can be

---

1 This chapter is partially based on a paper "Exploratory work on the application of distributed strain monitoring for submarine power cables" by S. Cherukupalli, F. Kammann, L. Molimbia, and S. Lima (2018). *CIGRE Science and Engineering Journal*, 2018;12:100–108.

used for measuring temperature and strain. Devices for locally resolved temperature and strain measurement employed in the tests described below make use of Brillouin scattering. Localization along the fiber length is then achieved via measurements of the signal travel time.

For the measurement to take place, only a FOC fixed on the measuring object without electric wiring or discrete electronic or optical sensors is needed. Fiber optics are standard telecommunication products and, therefore, fairly inexpensive and commercially available. Thereby, it would be relatively easy to attach a DSM system to an existing power cable with such fibers already installed. In addition, the measuring process is immune to electromagnetic interference. The measuring device operates autonomously with real-time alarm processing for the detection of critical events. The measurement data can also be transmitted to a database server for permanent storage and visualization. Temperature, strain, and event data are transmitted to the user process control system via various protocols. Application examples can be found as discussed in Chapter 5.

The distributed strain sensing system described below is based on a spontaneous Brillouin scattering in standard telecommunication fibers and does not require a fiber loop or special sensor elements. The instrument is cooled completely passively and does not contain any moving parts, offering a near maintenance-free solution, unlike current market systems that utilize cooling fans. The instrument is designed for use in industrial environments. With a measuring time below 10 minutes, the strain resolution is better than $20\,\mu\varepsilon$ for fiber distances of up to 50 km.

For the presented investigation, a spatial resolution of 1 m was used along the cable length with a strain range of up to $20\,000\,\mu\varepsilon$. The scan measurement window was only a few minutes in length.

In the following, we will offer brief review of construction of tight buffer tube cable followed by a presentation of the DSM equipment. Next, we will discuss the experimental test setup, results obtained, and how strain was calculated and compared using classical methods. Some discussion and recommendations are also offered.

11.3.2 The Importance of Tight Buffer Cable

FOCs may be either based on a loose tube or tight buffer design. In loose tube cables, the fiber itself can move freely and is always longer by a certain factor than the surrounding buffer. In this way the FOC can accommodate certain mechanical forces before the fiber is deformed. Tight buffer FOCs are intentionally produced with the fiber in fixed contact with the buffer. In this way all forces acting on the outside of the cable are also transferred up to a certain percentage to the fiber. Hence, tight buffer FOCs are the design of choice for most strain sensing applications.

11.3.3   Description of the Brillouin Optical Time Domain Reflectometer (BOTDR) System for Strain Measurement

The measurement performance was evaluated following the International Electrotechnical Commission (IEC) Standard 61757 definitions for DTS. The temperature repeatability and spatial temperature resolution amount to about 1 °C at 50 km distance. This corresponds to a measurement time below 10 minutes (see Figure 11.1) and a spatial resolution of 3 m. This performance can be useful to confirm the requirements for monitoring submarine power cables.

With a single-ended measurement system, there are significant advantages that make it very amenable for field application. For example, if a submarine cable includes an embedded fiber, by placing an instrument at the shore end, and connecting it to the accessible fiber, the cable can be monitored without worry, offering significant advantage.

11.3.4   Experimental Setup

Figure 11.2a shows the cross-section of the multimode loose tube cable and Figure 11.2b shows the single-core 25 kV class submarine cable with the steel armor and polypropylene yarn bedding. Figure 11.3 provides a description of the test setup where 24 m (76 ft) of distribution submarine cable was held with the help of Kellum grips and stretched to maintain some tension in the cable (see Figure 11.4). Single-mode optical fiber cables were spirally wrapped with an initial pitch of 1 m and subsequently with a pitch of 3 m on the same cable in

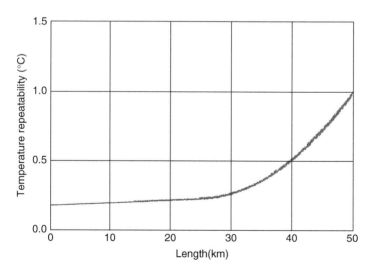

Figure 11.1 Temperature repeatability according to IEC 61757-3-1 along a 50 km fiber in a lab setup ($T = 20$ °C, measurement time 6.5 minutes).

(a)

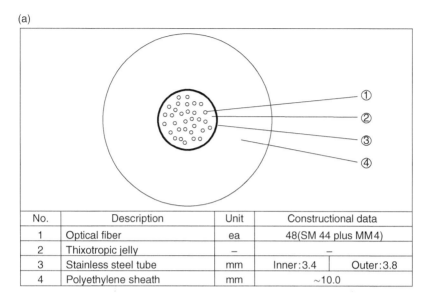

No.	Description	Unit	Constructional data	
1	Optical fiber	ea	48(SM 44 plus MM 4)	
2	Thixotropic jelly	–	–	
3	Stainless steel tube	mm	Inner : 3.4	Outer : 3.8
4	Polyethylene sheath	mm	~10.0	

(b)

Figure 11.2 (a) Cross-section of the 48-fiber count as well as (b) the tested single core submarine cable. (*Source:* courtesy of BC Hydro.)

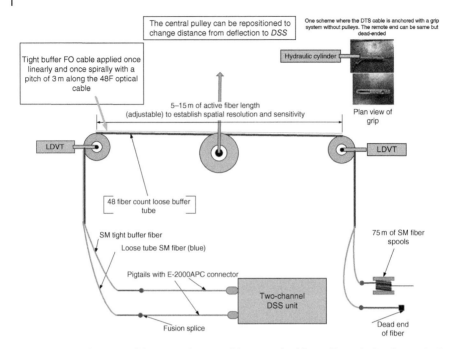

Figure 11.3 Schematic of the general setup of the optical cable and how the load is applied at two different locations.

turn and measurements were undertaken to assess the sensitivity to the pitch. To allow easy comparison between the DSM system and the classical method of measuring strains, strain gauges were also applied on the polypropylene yarn of the submarine cable, on the armor (by removing a few layers of the polypropylene yarn), and at the two loading sites to help correlate the optical and mechanical strains. Figure 11.5 shows a close-up view of these sites. These mechanical strain gauges were then connected to a data logger which was capable of recording and continuously monitoring changes of strain with time for the various loads applied gradually and rapidly. The continuous strain changes with respect to time for these two loading conditions were also recorded.

Once the tests were completed, the two FOCs were unraveled (unwrapped) and then reapplied on the distribution submarine cable linearly along the axis of the cables. The purpose of this change was to determine the sensitivity of a spirally wound optical fiber versus a fiber that is coaxial with the cable.

Load tests with this linear attachment were repeated with loading at two different locations along the cable. To avoid damage to the optical fiber, an empty wooden cable reel was attached to a spreader bar (~8 m from the instrument) and connected to the hoist system to lift the cable. This created a bow in the cable resulting in a measureable strain. The position of this reel was

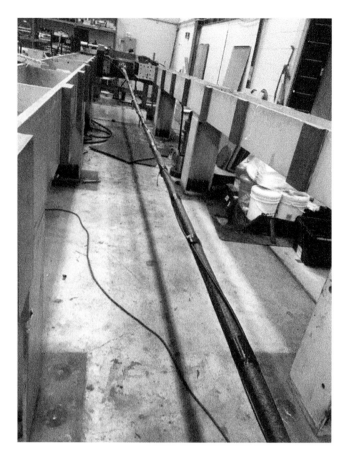

Figure 11.4 Submarine cable placed in the long bed anchored with Kellum grip and the two fiber optic cables spiraled at a 3 m pitch and subject to loading.

carefully monitored at the end of each loading cycle and recorded at three different loads (deflections) and repeated twice. The empty wooden reel was then moved 5 m farther along cable, making it about 15 m from the instrument. The same deflection or load was applied using the wooden reel at this new location and the strain was monitored.

The optical fiber used for the test had the characteristics described in Table 11.1.

11.3.5 Measurement Results

Figures 11.6 and 11.7 show the measured strain recorded by the DSM system for the sensor that was applied linearly and then spirally along the cable axis. It is evident that each condition yielded different results. The computed mechanical strain at the load location is overlaid as "dots" on the same plot.

Figure 11.5 Mechanical strain gauges applied in the armor and the polypropylene yarn to monitor strain using conventional techniques.

Table 11.1 Physical parameters of tight buffer FO cable.

Description	Value
Crush strength of the tight buffer cable	Not provided
Desired active length	Adjustable at 5, 10, and 15 m
Required length of fiber lengths at ends	50 m each
Minimum bend radius of tight buffer cable	63 mm (2.5″)
Required pitch to be applied on the tight buffer cable	3 m
Maximum tensile strength of tight buffer cable (long term)	400 N
Maximum tensile strength of tight buffer cable (short term)	1350 N (30–60 min)
Maximum allowable strain on the tight buffer FO cable (long term)	0.4%
Maximum allowable strain on the tight buffer FO cable (short term)	1%
Strain resolution of the DSM system	20 $\mu\varepsilon$

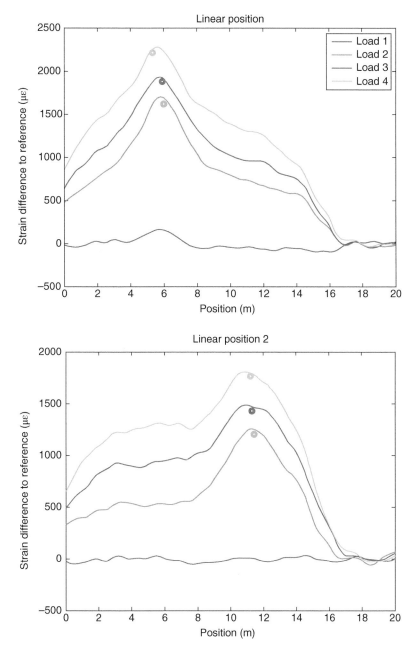

Figure 11.6 Strain recorded when the cable was subjected to different loads and at two different locations, near and farther from the DSM instrument. In this case the tight buffer tube cable has been taped coaxially along the submarine cable.

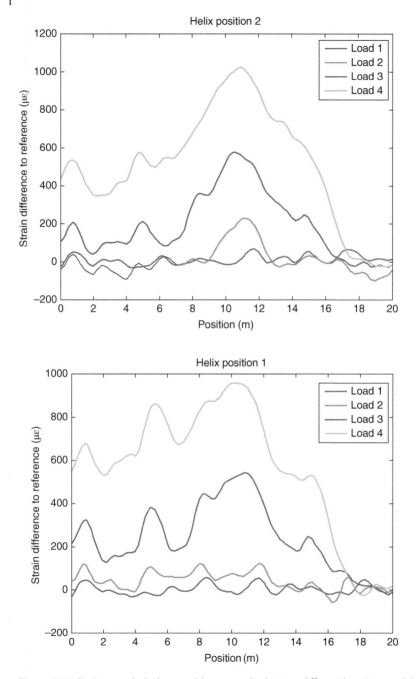

Figure 11.7 Strain recorded when a cable was applied at two different locations and the load applied at two different physical positions that were near and farther from the DSM instrument. In this case the tight buffer tube cable has been wound spirally along the test cable.

As expected, the peak strain along the cable occurred at the physical location where the load was applied. Figure 11.7 shows similar optical strain with the same test procedure. In this case the tight buffer sensing fiber is wrapped helically around the submarine cable axis. The strain shown in Figure 11.7 is characterized by a more oscillatory nature. This means that the strain is felt by the tight buffer all along the cable. The ability to distinguish between loads applied at the near-end and far-end locations is lost.

11.3.6 Discussion

The DSM system worked very well and was very sensitive in detecting strain on the cable that was not picked up by the conventional sensors (see Figures 11.8 and 11.9). Results with a tight buffer optical cable applied linearly along the cable was the most promising installation style.

Results with the loose tube 48 fiber-count cable were not very promising in the present setup. The reason for this shortcoming may be one of the following options:

- The loads applied were too small to cause a "stretch" in the loose tube FO cable; hence, the DSM did not see this strain.
- This is not an unexpected result as the FOCs are produced with excess fiber lengths of some 0.2%.
- Higher loads on the system were not applied as there was concern about the integrity of the "Kellum grips" and likelihood of causing slippage between them and the cable.

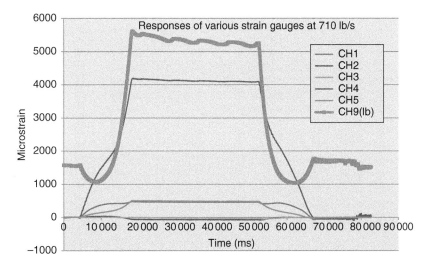

Figure 11.8 Strain recorded using conventional strain gauges when the load was applied rapidly. The channel numbers (CH1, CH2, …) represent the identification of the location of the strain gauges applied at different locations along the cable. CH9 is the load.

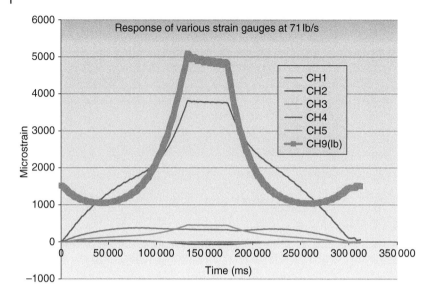

Figure 11.9 Strain recorded using conventional strain gauges when the load was applied slowly. The channel numbers (CH1, CH2, …) represent the identification of the various strain gauges applied along the cable.

Spiraling the sensing FOC showed several strain locations along the cable with the most pronounced signal at the loading site. In the helical fiber installation, the cable experiences tension on the one side and compression on the other when the power cable is bent which might explain this oscillatory response.

The Brillouin interrogator had a certain spatial resolution. For a helically installed FOC, the response depends on its pitch and the diameter of the power cable. The helically wrapped sensing cable now sees just a part of that deformation as it is at a certain angle to the bending direction. Additionally, a certain length of cable will always see a mix of strain information in that length. For example, for a spatial resolution of some meters and a pitch of some 10 cm, in the spectrum the interrogator sees is a mix of compressed and strained parts.

Referring to Figure 11.9, we can observe that the optical strain system was considerably more sensitive than the conventional strain sensor. The tensile load applied on the cable structure changes (due to the spiral armor) which can result in the strain responses being different along the cable.

When the spatial resolution is chosen reasonably better than the pitch, certain stress can be calculated by using a factor to account for the helical installation.

Figure 11.7 shows the results for 3 m pitch with a spatial resolution of 1 m. For both positions, the strain over the whole length increases when bending the cable. This is overlaid by a pattern that is expected from the helical installation, e.g., for position 1 the strain for load 3 at 5 m is around 400 µε while the close-by position 3.5 m shows around 200 µε. This is because the positions of highest compression are approximately 1.5 m away from the positions of highest tension for a pitch of 3 m and a relatively small diameter. However, this pattern cannot be seen continuously along the cable. This indicates that the expected strain transfer along the cable is of changing quality. Improving the attachment of the FOC to the power cable is expected to lead to clearer results.

There were differences in time response of the conventional sensors to slow and rapid load applications. Improved understanding is required to determine the transient response of the optical system to shock loads and how two strains recorded can be translated to cable-acceptable tension violations which are extremely important during submarine cable laying operations.

In a separate test, the authors have implemented a dynamic measurement mode which enables acquisition rates of approximately 3 Hz.

11.4 Calculation of the Cable Stress from the Strain Values

Cable stress can be calculated from the strain values within certain limits. The force acting on the outside of the FOC is transferred layer by layer always by a different percentage until it reaches the fiber, which is the sensor. A straightforward way to account for this is to characterize the FOC on a pulling bench where the force acting on the outside and the response of the sensor is measured. With this characterization, the stress can be calculated by knowing the strain values of one or more FOCs along the cable. While for longitudinal stress only one FOC is sufficient, for more complex loads such as local bending more than one FOC is needed.

To determine the dynamic response of submarine cables, the vessel response characteristics need to be coupled with the hydrodynamic loading of the cable. Using dynamic analysis software for offshore marine systems, such as OrcaFlex or Flexcom, the calculated vessel motions are applied as an external loading at the cable hang-off point. The resulting effective tension, T_e, calculated along the cable is determined from the following:

$$T_e = T_w + P_o A_o - P_i A_i \tag{11.1}$$

where

T_w = wall tension (N)
P_o/P_i = external and internal pressure, respectively (N/m^2)
A_o/A_i = external and internal cross-sectional stress areas, respectively (m^2)

and

$$T_w = EA\varepsilon - 2v\left(P_o A_o - P_i A_i\right) + \frac{K_t \tau}{L_0} + EAC\frac{dL/dt}{L_0} \tag{11.2}$$

where

EA = axial stiffness of the cable (Young's modulus × cross-sectional area of cable) (m^2)
ε = total mean axial strain
L = instantaneous length of segment (m)
L_0 = unscratched length of segment (m)
v = Poisson ratio
K_t = tension/torque coupling
τ = segment twist angle (in radians)
C = damping coefficient
dL/dt = rate of increase of the length (m/s)

Neglecting torsion and damping effects, the effective tension can be directly compared with the total mean strain the cable is subjected to. For the tight buffer arrangement, the elongation of cable and fiber is closely linked.

Hence, for the test setup discussed in Section 11.3.5, the estimated strain values from pushing cable laterally, as shown in Figure 11.8, can be directly compared to the fiber strain results.

11.5 Conclusions from the DSM Tests

Early work on establishing the feasibility of using a tight buffer optical cable has shown that it is quite capable of detecting strain on a power cable subjected to normal installation loads. Within the limitations of the test setup it was shown that when the load was moved by 5 m, the DSM system was capable of identifying where the load was being applied along the 15 m sample.

The DSM system is highly sensitive to subtle changes in the cable when subjected to tensile or compressive loads. The spatial resolution and sensitivity are much higher than conventional technology. Further work is needed to relate strain to cable-allowable tensions which will be a function of the tight buffer tube cable, cable construction, and so on. Installed on a submarine cable, this would, therefore, allow for precise determination of the touchdown location. The ability to detect distributed strain by itself is a very important finding.

11.6   Distributed Acoustic Sensing

Distributed acoustic sensing (DAS) systems are fiber optic-based optoelectronic instruments which measure acoustic interactions along the length of a FOC. Similar to the DTS systems, it is unique in that it provides a continuous (or distributed) acoustic profile along the length of the FOC and not at discrete sensing points.

When a pulse of light travels down an optical fiber, a small amount of the light is naturally backscattered through Rayleigh, Brillouin, and Raman scattering (as described in Chapter 1) and returns to the sensor unit. The nature of this scattered light is affected by tiny strain events within the optical fiber structure, which themselves are determined by the localized acoustic or seismic environment. By recording the returning signal against time, a measurement of the acoustic field all along the fiber can be determined. Some DAS systems commercially available have frequencies ranging from millihertz to hundreds of kilohertz.

Thus, DAS is a sophisticated variant of an optical time domain reflectometer (OTDR) that monitors the coherent Rayleigh backscatter noise signature in a FOC as pulsed light is sent into the fiber. The coherent Rayleigh noise generates fine structure in the backscatter signature of the fiber cable. The DAS focuses on the Rayleigh component to increase its prominence in the backscatter trace.

These units are generally optimized to measure small changes in the coherent Rayleigh noise structure that occurs from pulse to pulse. Since the coherent Rayleigh noise structure is generated interferometrically within the fiber by the relative locations and strengths of local scattering centers intrinsic to the structure of the glass, very small physical (acoustic or vibration) disturbances at a point in the fiber can make detectable changes in the interferometric signal.

DAS technology uses a standard telecom FOC and offers the flexibility to operate on single-mode or multimode fiber without the introduction of any external or additional apparatus, with no loss of signal quality while preserving the true acoustic nature of the measurement. This unique feature makes it possible to access legacy fiber optical installations that were installed for DTS for new acoustic surveys, for example, borehole seismic or flow measurements. These optical fibers can be installed as an asset is being built, with DAS in mind. Therefore, by default the "sensor" is immune to EM or RF interface – is also inert and requires no power along the entire length beyond the DAS interrogator. These DAS capabilities are in use in the oil, gas, and border protection businesses. Any single-mode fiber can quickly be turned into a series of listening devices using DAS with minimal fiber work at either end of the monitored fiber section. Using spare capacity on the existing fiber cables next to a track makes it possible for similar and related applications to be realized for power cable-specific requirements.

Specialized fibers are only required when the operation is to be done at elevated temperatures (>100°C). With today's technology, the range of DAS systems is typically up to 50 km per sensing fiber.

DAS is a very versatile and evolving technology. Some of the more common areas of application that we see today include the following:

- Land cable monitoring
- Submarine cable monitoring
- Perimeter intrusion detection
- Fire detection

The spatial resolution is mainly determined by the duration of the transmitted pulse, with a 100 ns pulse giving 10 m resolution being a typical value. The amount of reflected light is proportional to the pulse length so there is a trade-off between spatial resolution and maximum range. To improve the maximum range, it would be desirable to use a longer pulse length to increase the reflected light level but this leads to a larger spatial resolution. Typically, most systems have a spatial resolution of 5–10 m.

There are a number of other distributed fiber sensing techniques that rely on different scattering mechanisms and can be used to measure other parameters.

Due to the large amounts of data produced by DAS systems, it is critical that there is an appropriate strategy for data management, processing, and visualization. These systems acquire data at up to 20 sensing points at a rate greater than 10 kHz. This equates to a rate where a terabyte drive can be filled in a matter of days! Typically, the interrogation units are networked to a processing unit (industrial PC or server) which manages the data storage and processing. Usually, there is a rolling buffer for storing the raw data as it is seldom practical to store more than this.

The processing unit is programmed with a range of smart algorithms for interpreting the raw data and analyzing if this matches predefined events such as, for example, an intruder event. The fiber optic sensing cable will be divided into a number of zones in which specifically chosen algorithms will be selected and alarms allocated within each zone. There are a number of ways to visualize these events. One is to use visualization software which is specific to the unit which, for example, may show the path of the fiber optic against a site map or diagram and, in the event there is an occurrence, it will highlight the location of the event and show an alarm. An alternative is that the DAS software is interfaced with an existing third party SCADA, control, or security software package, in which case the event will be highlighted in the third party software. Figures 11.10 and 11.11 illustrate the possible displays.

In addition, as all sensing points are phase-matched, the acoustic response along the fiber can be combined to enhance the detection sensitivity by two orders of magnitude, which would enable to step beyond the performance of

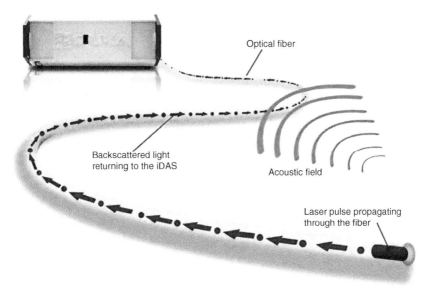

Optical fiber

Backscattered light
returning to the iDAS

Acoustic field

Laser pulse propagating
through the fiber

Figure 11.10 Schematic of a DAS system. (*Source:* schematic of courtesy from the https://spie.org/membership/spie-professional-magazine/listening-with-light?SSO=1.)

LIGHTLINE™ fiber optic sensing

Figure 11.11 Schematic of a DAS system exhibiting its ability to detect the presence of vehicles, humans, or animals in proximity to the sensing fiber optic cable. (*Source:* courtesy of the website: https://spie.org/news/spie-professional-magazine/2019-april/listening-with-light?SSO=1.)

current point sensors and, if desired, achieving highly directional information. These capabilities can be further enhanced by forming the sensing cable into an acoustic antenna whose sensitivity and frequency response can be adjusted dynamically.

The area that has seen a lot of growth has been in Oilfield Services where DAS systems offer real-time information to improve reservoir management and ultimate recovery. Such systems are used for borehole seismic processing and imaging, borehole seismic acquisition/VSP, hydraulic fracture profiling, and production flow monitoring.

Integrating distributed data generally is challenging since terabytes of data can be generated quite easily and quality control, analysis, and interpretation from many physical and production workflows can be demanding. Moreover, in this discipline integrating 3D data that spans depth, time, and measurement can present another obstacle, with many of today's conventional tools being capable of handling only two of these three data sets. Some vendors today offer software suites that allow easier integrations of such large distributed data sets.

11.7   Potential DAS Applications in the Power Cable Industry

With the advent of large offshore wind farms and the need to lay cables to collect and transmit the power to land, many AC and DC low- and high-voltage cables are being laid. While these systems are manufactured, installed, and operated with care, quite often cable failures have occurred. When such failure occur, the ability to detect, locate, and repair the failure to restore the power transmission has to be done quickly to prevent excessive revenue loss. Classical technique would have required TDR techniques or underwater remote-operated vehicle survey to locate the defects and these campaigns can be time consuming and expensive. This is especially true when the weather is rough. If an optical fiber is laid alongside these corridors, by placing it as an integral part of the power cable or alongside as an intrusion detection system similar to what is done in the pipeline or railway industry, a DAS system offers an elegant solution to detect and locate these events rapidly and successfully. In busy shipping lanes, the DAS-enabled optical fibers can be used as early warning systems to alert the shipping authorities or the Captain of the ship that is sailing in these waters about the presence of expensive assets underwater and avoid dropping anchor and causing damage.

In a study report in CIGRE TB 379 (2009) that recorded the service experience of high voltage underground submarine and land cables, it was reported that 34% of all faults in submarine cable installations were due to mechanical damage caused by third parties digging into cable systems. The DAS technology offers a remedy to provide advance warning of such impending mechanical activity in the proximity of both land and submarine cables. Alternatively, it can also provide clear evidence of the cause of such damaging activity once it has occurred. While many submarine cables are protected, over 50% of faults have occurred on unprotected cables. Buried cables are well protected against fishing gear but they can still be damaged by anchors. Statistics reported in this brochure has shown that 85% of "26 defined" faults were due to external influences and almost 50% of damage was known to be caused by anchors.

For merchant cable installations, these systems can help in ascertaining liability should such an unfortunate event occur as the DAS systems are capable, when properly tuned, to detect the presence of an approaching ship in its vicinity and the speed at which this ship/trawler is moving.

On shore, underground cable damage is commonly attributed to unintentional mechanical digging along the major road networks. They are also impacted by copper cable theft, another epidemic affecting utilities and communities worldwide. However, copper is not the only metal criminals target. Aluminum conductors installed on overhead transmission lines and steel sections from transmission towers are also being vandalized and sold as scrap metal. Although third-party interference and theft pose a serious threat to the continuity of power supply and public safety, they can be prevented. DAS systems provide the opportunity for online detection and location of third-party interference near the vicinity of power cables and towers, including a person walking up to 5 m, vehicle movement 10 m out, and digging up to 20 m.

A typical display of a DAS system is presented in the form of a water fall plot. This is a 2D plot as illustrated in Figures 11.12 and 11.13 that show acoustic activity as a function of time and distance from the monitoring site. This example illustrates the nature of activity detected along a FOC using DAS technology.

11.8 An Example of a DAS Application in the USA

The Hudson Transmission Project is a 660 MW electric transmission link between New York City and PJM Interconnection. Its main purpose is to provide a new source of electric power for the New York City customers of the New York Power Authority (NYPA). The consortium of Prysmian and Siemens was responsible for the design, supply, and installation of a 345 kV extra high-voltage alternate current (HVAC) land and submarine transmission line running along a total route of approximately 8 miles (about 13 km) to transfer 660 MW of existing power from the transmission grid in Ridgefield, New Jersey to New York City. The Hudson Project was deemed to be of strategic importance for the City of New York where energy load is constantly increasing. This project was to help replace resources that may be retired over the next several

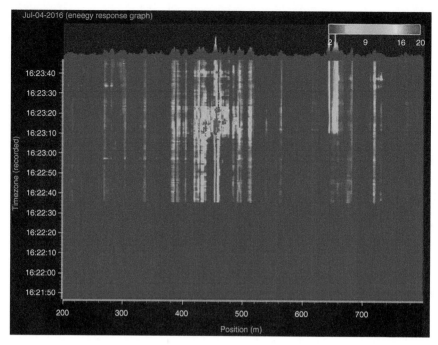

Figure 11.12 The *X*-axis represents the distance from the DAS system along the fiber, the *Y*-axis represents time and the color intensity of the activity. (*Source:* courtesy of AP Sensing.)

(a) (b) (c)

Manual digging Mechanical digging Leakages

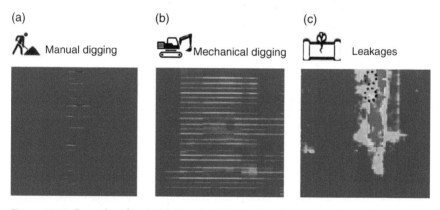

Figure 11.13 Examples of typical DAS water fall plots showing various types of activities such as (a) manual digging, (b) backhoe activity, and (c) pipe leakage. (*Source:* courtesy of AP Sensing.)

years as well as strengthen the overall reliability of the power supply system in NYC as a long-term infrastructure asset. It was also expected to provide New York City customers access to more diverse sources of power, including renewable supplies and natural gas.

The cable system connecting NJ to NY is 8-mile long with approximately 4 miles running beneath the Hudson River. The river section comprised a 2000 mm^2 copper conductor XLPE insulated 345 kV cables with two 36-count optical fibers integrated with the trefoil cable system. A DAS system based on coherent optical time domain reflectometry capable of 70 km range was used to monitor the 11.42 km of cable. The points of interest were identified to be where this cable enters the Hudson River on the New Jersey side and enters New York. Tap tests were performed to determine the exact position of these points of interest and impulsive noises were also generated with a hammer on the metal vault covers.

Figure 11.14 shows the results from the tap test and the impulsive noises generated on both fibers at two different distances (11.095 and 11.252 km).

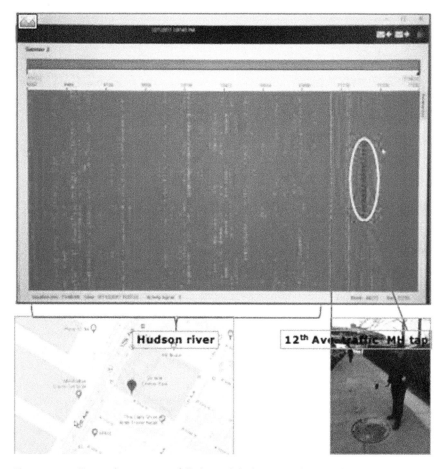

Figure 11.14 Figure shows a water fall plot and the location where the hammer test was conducted on a steel manhole cover as a test. The traffic noise may also be seen in the waterfall plot on the 12th Street. (*Source:* courtesy of Prysmian Cables.)

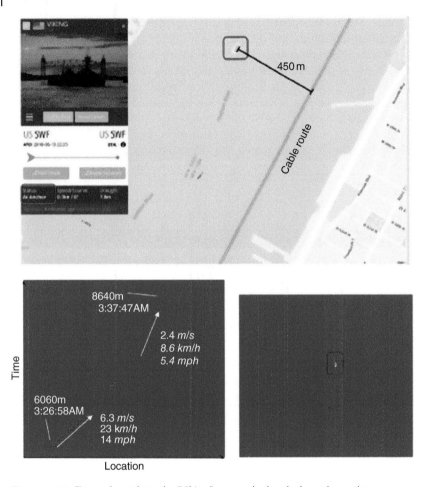

Figure 11.15 Figure shows how the "Viking" was tracked and where the anchor was dropped in the water fall plot. (*Source:* courtesy of Prysmian Cables.)

The water fall chart shown in this figure illustrates the results from the tap test. At the distance of about 11.05 km, there is a very noisy portion on the waterfall chart, which is caused by the traffic noise in New York City on 12th Avenue.

In addition, alarms were set up to detect ships/vessels approaching and anchoring near the submarine EHV cable. The "Viking" ship was monitored (which anchored) after moving half way down the Hudson River along Manhattan and can be seen from the DAS waterfall diagram as shown in Figure 11.15.

Prior to anchoring, vessel velocity is reduced and followed by a huge acoustic event at the end of the motion and the vessel decelerated rapidly and came to a full stop. Anchoring occurred 450 m away from the EHV submarine power cable (waterfall diagram shows acoustic event intensity depending on the distance).

11.9   An Example of a DAS Application in Scotland

In an online publication, the following case study is presented and described below.

> While operating on an offshore windfarm in Scotland, midway between the Galloway and Cumbrian coasts, our client experienced a short circuit on a main export cable approximately 15 km in length, of which 13 km was located subsea. Until the fault could be located and repaired, 80 MW of offshore power would be sitting idle.
>
> Existing Time Domain Reflectometry (TDR) methods using a high voltage pulse generator indicated the fault was 2 km offshore. A repair in this location would cost the operator £8M, in addition to placing the wind turbines offline for up to 6 months – accruing a potential loss of revenue of £90 000 per day. The TDR technique is a manual and time consuming process to locate a fault. Even when operating well the accuracy of the TDR method is low (±300 m on a 15 km cable).
>
> As with many export and inter-array subsea power cables, the export cable had an existing fibre optic cable integrated within the structure – meaning the use of Distributed Acoustical Sensing (DAS) was possible.
>
> OptaSense® DAS technology is a proven method for detecting any activity in the vicinity of an asset over long distances. DAS technology allowed the existing fibre optic cables into an array of virtual microphones with no additional infield equipment required. Using this solution an operator is able to detect, classify and locate any breakdown or threatening events near to the assets in real time. With the DAS cable already installed, OptaSense experts were able to mobilise quickly to access the site within 24 hours. In fact, the fault was located and verified within 4 hours of OptaSense arriving on site. The location of the discharge noise, which was caused by the generator, was determined using the OptaSense interrogator unit. This interrogator unit was able to detect the fault in minutes, to an accuracy of ±10 m. This accuracy allowed experts to determine the fault was actually located on land, eliminating up to eight weeks of repair time at £10K per day and lost revenue for the operator. By identifying an onshore fault we additionally eliminated the need for high cost vessels and ROV's. Even though the fibre was located 2 m away from the power cable, the OptaSense

interrogator unit offers a long range capability of 40 km that allowed the fault to be easily and quickly detected onshore.

(Reference: https://www.optasense.com/wp-content/uploads/2017/05/ Condition-Monitoring-for-Power-Cables_Case-Study_A4-Digital.pdf; Figure 11.16)

11.10 Conclusions

In summary, examples show how power utilities sector can benefit from DAS technology in various applications:

- Short circuit detection and localization – identification of the fault such as short circuit and locate within 5–10 m of occurrence for subsea and buried cables.

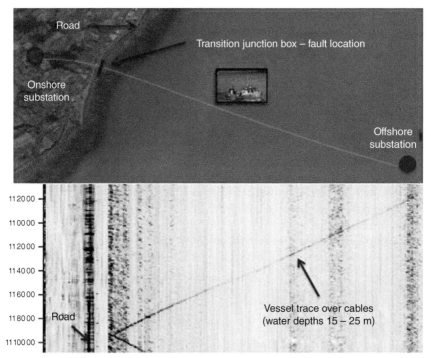

Trace of vessel with the water depths between 15 and 25 m at high tide

Figure 11.16 Overview of a power cable with integrated fiber and the DAS response showing the presence of a vessel in close proximity to the power cables. (*Source:* courtesy from the https://www.optasense.com/wp-content/uploads/2017/05/Condition-Monitoring-for-Power-Cables_Case-Study_A4-Digital.pdf.)

- Protection against third party intervention (TPI) for buried onshore and off-shore cables.
- Early detection with the ability to monitor in real time to detect, locate, and alert. This provides the opportunity to intercept before an event occurs to safeguard the asset, such as the early detection of anchor drags, fishing nets, etc.
- Detect vortex-induced vibrations and free scan in a cable (0.25–2.5 Hz).
- Long measuring ranges with capabilities extending beyond 40 km.

Bibliography

Anders, G.J. (1997). *Rating of Electric Power Cables. Ampacity Calculations for Transmission, Distribution and Industrial Applications.* New York: IEEE Press/ McGraw Hill.

Anders, G.J. (2005). *Rating of Electric Power Cables in Unfavorable Thermal Environment.* New York: IEEE Press/Wiley.

Anders, G.J., Napieralski, A., Zubert, M., and Orlikowski, M. (2003). Advanced modeling techniques for dynamic feeder rating systems. *IEEE Transactions on Industry Applications* 39 (3): 619–626.

Avila, A. and Vogelsang, R. (2010). Experiences in manufacturing, testing, installing and operating of 500kV cable systems including temperature sensing and PD monitoring. *Paper B1-103, CIGRE Paris Session 2010.*

Balog, G., Nerby, T.I., and Kaldhussaeter, E. (1999). Cable temperature monitoring. *Paper B8.4, 5th International Conference on Insulated Power Cables, JiCable'99.*

Bao, X., Dhliwayo, J., Heron, N. et al. (1995). Experimental and theoretical studies on a distributed temperature sensor based on Brillouin scattering. *Journal of Lightwave Technology* 13: 1340–1346.

Bosselmann, T. (1997). Magneto- and electro-optic transformers meet expectations of power industry. *Proc. OFS'12,* pp. 111–114.

Brakelmann, H., Hirsch, H., Rohrich, A. et al. (2007). Adaptive monitoring program for dynamic thermal rating of power cables. *Proceedings of Jicable'07,* Versailles, France, 2007.

Cantergrid, B., Bayarad, C., and Trzeguet, J.P. (1987). Composite power and optical submarine cable. *Paper B9.3, JiCable'87.*

Carslaw, H.S. and Jaeger, J.C. (1959). *Conduction of Heat in Solids,* 81–83. Oxford: Clarendon Press.

Cherukupalli, S. (2006). Application of distributed fiber optic temperature sensing on BC Hydro's 525kV submarine cable system. *Paper B1-203, CIGRE Paris Session 2006.*

Distributed Fiber Sensing and Dynamic Rating of Power Cables, First Edition.
Sudhakar Cherukupalli and George J. Anders.
© 2020 by The Institute of Electrical and Electronics Engineers, Inc.
Published 2020 by John Wiley & Sons, Inc.

Cherukupalli, S. and MacPhail G.A. (2002). Application of Fiber-Optic Distributed Temperature Sensing to Power Transmission Cables at BC Hydro. *EPRI Tech. Rep. 1000443* (May 2002).

Cherukupalli, S., Korczynski, J., and MacPhail, G.A. (1996). A sequential approach to increasing underground transmission line ampacities at BC Hydro. *EPRI Conference on the Future of Power Delivery,* Washington, DC (April 1996).

Cherukupalli, S., Adapa, R., and Bascom, R. (2018). Implementation of quasi-real-time rating software to monitor 525kV cable systems. *IEEE Transactions on Power Delivery.*

Chimi, E., Heimbach, H.B., Bader, J., and Luternauer, H. (2014). Pilot project for cable capacity monitoring using optic fiber cables and prognostic software in the city of Zurich. *Paper Number B1-208, CIGRE Paris Session 2014.*

CIGRE. (2015). Recommendations for mechanical testing of submarine cables. *CIGRE WG B1.43* (June 2015)

CIGRE TB 379. (2009). Update of service experience of HV underground and submarine cable systems. *CIGRE Technical Brochure 379*, p. 86.

CIGRE WG B1.02. (2004). Optimization of power transmission capability of underground cable systems using thermal monitoring. *CIGRE Technical Brochure 247* (April 2004).

Culshaw, B. and Dakin, J.P. (eds.) (1996). *Optical Fiber Sensors*, vol. 3 and 4. Boston: Artech House.

Day, G.W., Deeter, M.N., and Rose, A.H. (1992). Faraday effect sensors: a review of recent progress. *Proc. SPIE, PM07*, pp. 11–26.

De Wild, F., Boone, W., Geest, H.V., and Smit, J. (2007). Dynamic ratings systems in general and hybrid 150kV transmission system. *Paper A6.5, 7th International Conference on Insulated Power Cables, Ji Cable'07, Versailles, France.*

Diaz-Aguiló, M., de León, F., Jazebi, F., and Terracciano, M. (2014). Ladder-type soil model for dynamic thermal rating of underground power cables. *PETSJ* 1 (1): 21–30.

Domingues, I.T. (2010). Development of one specialist system to determine the dynamic current capacity of underground transmission lines with XLPE cables. *Paper B1-202, CIGRE Paris Session 2010.*

Donazzi, F. and Gaspari, R. (1998). Method for the management of power cable links. *Cigre 21-203.*

Duchateauf, F. (1987). Optical fibers in power underground links. *Paper B9.1, JiCable'87.*

Farahani, A., Anders, G.J., Bic, E., and Keppler, U. (2015). A new approach for estimation of the dynamic thermal rating model parameters based on the IEC standard. *Proc. Jicable'15*, Versailles, France.

Ferdinand, P., Ferragu, O., Lechien, J.L. et al. (1995). Mine operating accurate stability control with optical fibre sensing and Bragg grating technology: the BRITE-EURAM STABILOS project. *Journal of Lightwave Technology* 13: 1303–1313.

Foote, P.D. (1994). Fibre Bragg grating strain sensors for aerospace smart structures. *Proc. SPIE, 2361*, pp. 162–166.

Friebele, E.J., Askins, C.G., Putnam, M.A. et al. (1994). Distributed strain sensing with fiber Bragg grating arrays embedded in CRTM composites. *Electronics Letters* 30: 1783–1784.

Giussani, A., Lanfranconi, G.M., and Occhini, E. (1987). Submarine composite cables for power and data transmission – design philosophy. *Paper B9.4, 4th International Conference on Insulated Power Cables JiCable'87*.

Grattan, K.T.V. and Meggitt, B.T. (eds.) (1995). *Optical Fiber Sensor Technology*, 269–310. London: Chapman and Hall.

Grattan, K.T.V. and Meggitt, B.T. (eds.) (1998). *Optical Fiber Sensor Technology II: Devices and Technology*. London: Chapman and Hall.

Hammon, T.E. and Stokes, A.D. (1996). Optical fibre bragg grating temperature sensor measurements in an electrical power transformer using a temperature compensated fibre Bragg grating as a reference. *Proceedings of the 11th International Conference on Optical Fiber Sensors*, Sapporo, Japan, pp. 566–569.

Henderson, P.J., Fisher, N.E., and Jackson, D.A. (1997). Current metering using fibre-grating based interrogation of a conventional current transformer. *Proceedings of the 12th International Confernece on Optical Fiber Sensors*, Williamsburg, VA, pp. 186–189.

Hennuy, B., Leemans, P., Mampaey, B. et al. (2014). Belgian philosophy and experience with temperature monitoring of cable systems by means of distributed temperature sensing techniques and future PD monitoring techniques. *Paper B1-203, CIGRE 2014*.

IEC. (1989). Calculation of the cyclic and emergency current rating of cables. *International Electrotechnical Commission Standard Series 60853* (July 1989).

IEC. (2014). Electric cables – calculation of the current rating – current rating equations (100% load factor) and calculation of losses. *International Electrotechnical Commission Standard Series* 60287-1 (December 2006).

IEC. (2016). Fibre optic sensors Part 2-2: temperature measurement – distributed sensing. *International Electrotechnical Commission, Standard 61757-2-2*.

IEEE Standards. (2012). Guide for temperature monitoring of cable systems. *IEEE Standard 1718*.

Igi, T., Komeda, H., Mashio, S. et al. (2011). Study of the dynamic rating of a 138kV XLPE cable system by optical fiber monitoring. *Paper C9.4, 8th International Conference on Insulated Power Cables, JiCable'11*.

ITU Publication G.651.1. (2016). Characteristics of a 50/125 μm multimode graded index optical fibre cable for the optical access network.

Jacobsen, E. and Nielsen, J.F. (2012). Operational experience of dynamic cable rating. *Paper B1-205, CIGRE Paris Session 2012*.

Jacobsen, E., Nielsen, J.F., Salvin, S. et al. (2011). Dynamic ratings of transmission cables. *Paper C9.2, 8th International Conference on Insulated Power Cables, JiCable'11*.

Jacobsen, E., Nielsen, J.F., and Nielsen, S.B. (2012). Dynamic rating of transmission cables. *Paper B1-101, CIGRE Paris Session 2012.*

Jaeger, N.A.F., Polovick, G., Kato, H., and Cherukupalli, S.E. (1998). On-line dissipation factor monitoring of high voltage current transformers. *CIGRE, Paris* (July 1998).

Jones, J.D.C. (1997). Review of fibre sensor technique for temperature-strain discrimination. *Proceedings of the 12th International Conference on Optical Fiber Sensors*, Williamsburg, VA, pp. 36–39.

Kim, J., Cho, J., Lee, L., and Choi, T. (2011). Intelligent DTS and PD monitoring for underground distribution network. *Paper B8-2, 8th International Conference on Insulated Power Cables, JiCable'11.*

Kluth, R., Svoma, R., and Singh, K. (2007). Application of temperature sensing and dynamic strain monitoring to sub-sea cable technology. *Paper A9-6, 7th International Conference on Insulated Power Cables, JiCable'07.*

Kusuda, T. and Achenbach, P.R. (1965). *Earth Temperature and Thermal Diffusivity at Selected Stations in the United States.* Tech. Rep. NBS Rep. 8972. Washington, DC: U.S. Department of Commerce/National Bureau of Standards.

Leduc, J., Beauchemin, R., Chaaban, M., and Choquette, M. (2003). Monitoring and dynamic rating of 120 kV XLPE insulated cable circuits at Hydro-Quebec. *Paper C8.1.4, 6th International Conference on Insulated Power Cables, JiCable'03.*

Lee, S.K., Nam, S.H., Hong, J.Y. et al. (2003). Real-time ampacity estimation system for 345 kV transmission cable installed in tunnel. *Paper C8.1.5, 6th International Conference on Insulated Power Cables, JiCable'03.*

Li, H.J., Tan, K.C., and Su, Q. (2006). Assessment of underground cable ratings based on distributed temperature sensing. *IEEE Transactions on Power Delivery* 21 (4): 1763–1769.

Liu, Q. and Wang, W. (2011). Research on temperature precision influenced by fiber loss based on distributed temperature measurement system in cable monitoring. *Paper C7.2, 8th International Conference on Insulated Power Cables, JiCable'11.*

Liu, T., Fernando, G.F., Zhang, L. et al. (1997). Simultaneous strain and temperature measurement using a combined fiber Bragg grating/extrinsic Fabry-Perot sensor. *Proceedings of the 12th International Conference on Optical Fiber Sensors*, Williamsburg, VA, pp. 40–43.

Luton, M.H., Anders, G.J., Braun, J.M. et al. (2003). Real time monitoring of power cables by fiber optic technologies tests, applications and outlook. *Paper A1.6, 6th International Conference on Insulated Power Cables, JiCable'03.*

Marsh, T., Beer, L.J., and Shackleton, D.J. (1987). Submarine power cable incorporating telecommunications data and control links. *Paper B9.2, 4th International Conference on Insulated Power Cables, Jicable'87*, Versailles, France.

Marx, B., Rath, A., Kolm, F. et al. (2016). *Brillouin Distributed Temperature Sensing System for Monitoring of Submarine Export Cables of Off-shore Wind farms*, Sixth European Workshop on Optical Fiber Sensors, edited by Elfed Lewis, *Proc. of SPIE*, vol. 9916.

Measures, R.M. (1989). Smart structures in nerves of glass. *Progress in Aerospace Sciences* 26: 289.

Measures, R.M., Alavie, A.T., Maaskant, R. et al. (1994). Bragg grating structural sensing system for bridge monitoring. *Proc. SPIE, 2294*, pp. 53–60

Measures, R.M., Maaskant, R., Alaview, T. et al. (1997). Fiber-optic Bragg gratings for bridge monitoring. *Cement and Concrete Composites* 9: 21–23.

Mendez, A., Morse, T.E., and Mendez, E. (1989). Applications of embedded optical fibre sensors in reinforced concrete buildings and structures. *Proc. SPIE, 1170*, pp. 60–69.

Millar, R.J. and Lehtonen, M. (2006). A robust framework for cable rating and temperature monitoring. *IEEE Transactions on Power Delivery* 21 (1): 313–321.

Morey, W.W., Meltz, G., and Glenn, W.H. (1989). Fiber optic Bragg grating sensors. *Proc. SPIE, 1169*, pp. 98–107.

Neher, J.H. (1964). The transient temperature rise of buried power cable systems. *IEEE Transactions on Magnetics* PAS-83: 102–111.

Ogawa, Y., Iwasaki, J. I., and Nakamura, K. (1997). A multiplexing load monitoring system of power transmission lines using fibre Bragg grating. *Proceedings of the 12th International Conference on Optical Fiber Sensors*, Williamsburg, VA, pp. 468–471.

Olsen, R., Anders, G.J., Holboell, J., and Gudmundsdóttir, U.S. (2013). Modelling of dynamic transmission cable temperature considering soil specific heat, thermal resistivity and precipitation. *IEEE Transactions on Power Delivery* 28 (3): 1909–1917.

Popovac, A. and Damlianovic, P. (2006). Thermal monitoring of high voltage cables. *Paper B1-210, CIGRE Paris Session 2006.*

Pragnell, B., Gaspari, R., and Larsen, S.T. (1999). Real time thermal rating of HV cables. *Paper B8.2, 5th International Conference on Insulated Power Cables, JiCable'99.*

Rao, Y.J. (1997). Review article: in-fibre Bragg grating sensors. *Measurement Science and Technology* 8: 355–375.

Rao, Y.J. and Jackson, D.A. (1996). Review article: recent progress in fiber-optic low-coherence interferometry. *Measurement Science and Technology* 7: 981–999.

Rao, Y.J., Henderson, P.J., Fischer, N.E. et al. (1998). Wavelength-division-multiplexed in-fibre Bragg grating Fabry–Perot sensor system for quasi-distributed current measurement. *Proceedings of the Annual Conference on Applied Optics and Optoelectronics (IOP)*, Brighton, pp. 99–104.

Rogers, A.J. (1988). Optical-fibre current measurement. *International Journal of Optoelectronics* 3: 391–407.

Sakata, M. and Iwamoto, S. (1996). Genetic algorithm based real-time rating for short-time thermal capacity of duct installed power cables. *Proc. Intell. Syst. Appl. to Power Syst. Int. Conf., 1996*, pp. 85–90.

Schmale, M. and Drager, H.J. (2010). Implementation and operation of a cable monitoring system in order to increase the ampacity of a 220-kV underground power cable. *Paper B1-113, CIGRE Paris Session 2010*.

Schmale, M., Puffer, R., Glombitza, U., and Hoff, H. (2011). Online ampacity determination of a 220kV cable using an optical fiber based monitoring system. *Paper C9.3, 8th International Conference on Insulated Power Cables, JiCable'11*.

SEEFOM-MSP-01Document. (2016). Measurement specification for distributed temperature sensing (January 2016).

Smit, J.C. and de Wild, F.H. (2006). A dynamic rating system for an existing 150kV power connection consisting of an overhead line and an underground power cables. *Paper B1-305, CIGRE Paris Session 2006*.

Su, Q., Li, H.J., and Tan, K.C. (2003). Determination of cable dynamic rating from distributed temperature sensing and hotspot temperature analysis. *Paper A9.5, 6th International Conference on Insulated Power Cables, JiCable'03*.

Thevenaz, L., Facchini, M., Fellay, A. et al. (1999). Monitoring of large structures using distributed Brillouin fiber sensing. *Proc. OFS'13*, pp. 345–348.

Townsend, R.D. and Taylor, N.H. (1996). Fiber optic monitoring of temperature and strain along insulated pipework at high temperature. *Proc. Euromaintenance'96*.

Trimble, B. (1993). Fifty thousand pressure sensors per year: a successful fibre sensor for medical applications. *Proc. OFS'9*, pp. 457–462.

Udd, U. (ed.) (1995). *Fiber-Optic Smart Structures*. New York: Wiley.

Wald, D. (2012). Correlation between calculated transmission capacity and actual one. *Paper Number B1-214, CIGRE Paris Session 2012*.

Watanabe, T., Muramatsu, Y., and Nakanishi, M. (2015). Operating records and recent technology of DTS System and Dynamic Rating System (DRS). *Paper A3.6, Jicable'15*, Versailles, France.

Yan, Y., Su, X., and Xiao, C. (2011). The necessity analyses of distributed fiber-optic temperature monitoring by Xiamen power cable alarm case study. *Paper C7, 8th International Conference on Insulated Power Cables, JiCable'11*.

Index

Distributed Fiber Sensing and Dynamic Rating of Power Cables, First Edition.
Sudhakar Cherukupalli and George J. Anders.
© 2020 by The Institute of Electrical and Electronics Engineers, Inc.
Published 2020 by John Wiley & Sons, Inc.

IEEE Press Series on Power Engineering

Series Editor: M. E. El-Hawary, Dalhousie University, Halifax, Nova Scotia, Canada

The mission of IEEE Press Series on Power Engineering is to publish leading- edge books that cover the broad spectrum of current and forward-looking technologies in this fast-moving area. The series attracts highly acclaimed authors from industry/academia to provide accessible coverage of current and emerging topics in power engineering and allied fields. Our target audience includes the power engineering professional who is interested in enhancing their knowledge and perspective in their areas of interest.

Distributed Fiber Sensing and Dynamic Rating of Power Cables, First Edition.
Sudhakar Cherukupalli and George J. Anders.
© 2020 by The Institute of Electrical and Electronics Engineers, Inc.
Published 2020 by John Wiley & Sons, Inc.